METHODS IN MOLECULAR BIOLOGY

For further volumes:
http://www.springer.com/series/7651

For over 35 years, biological scientists have come to rely on the research protocols and methodologies in the critically acclaimed *Methods in Molecular Biology* series. The series was the first to introduce the step-by-step protocols approach that has become the standard in all biomedical protocol publishing. Each protocol is provided in readily-reproducible step-by-step fashion, opening with an introductory overview, a list of the materials and reagents needed to complete the experiment, and followed by a detailed procedure that is supported with a helpful notes section offering tips and tricks of the trade as well as troubleshooting advice. These hallmark features were introduced by series editor Dr. John Walker and constitute the key ingredient in each and every volume of the *Methods in Molecular Biology* series. Tested and trusted, comprehensive and reliable, all protocols from the series are indexed in Pub Med.

Microbial Transposon Mutagenesis

Protocols and Applications

Edited by

Steven C. Ricke

Department of Food Science, Center for Food Safety, University of Arkansas, Fayetteville, AR, USA

Cell and Molecular Biology Program, University of Arkansas, Fayetteville, AR, USA

Si Hong Park

Department of Food Science and Technology, Oregon State University, Corvallis, OR, USA

Morgan L. Davis

Department of Food Science, University of Arkansas, Fayetteville, AR, USA

Humana Press

Editors
Steven C. Ricke
Department of Food Science
Center for Food Safety
University of Arkansas
Fayetteville, AR, USA

Cell and Molecular Biology Program
University of Arkansas
Fayetteville, AR, USA

Si Hong Park
Department of Food Science and Technology
Oregon State University
Corvallis, OR, USA

Morgan L. Davis
Department of Food Science
University of Arkansas
Fayetteville, AR, USA

ISSN 1064-3745 ISSN 1940-6029 (electronic)
Methods in Molecular Biology
ISBN 978-1-4939-9569-1 ISBN 978-1-4939-9570-7 (eBook)
https://doi.org/10.1007/978-1-4939-9570-7

This Humana imprint is published by the registered company Springer Science+Business Media, LLC, part of Springer Nature.
The registered company address is: 233 Spring Street, New York, NY 10013, U.S.A.

Preface

Whole-genome sequencing and associated methodologies have greatly accelerated an understanding of the genetics of microorganisms. However, assigning functions to individual genes remains a challenge. Transposon mutagenesis represents an approach for identifying individual gene contributions to the phenotypic characteristics of a particular microorganism. Generation of transposon mutant libraries when coupled with whole-genome sequencing can pinpoint genes responsible for specific physiological functions associated with a microorganism. In *Microbial Transposon Mutagenesis: Protocols and Applications*, the first section of topics are categorized as methods for specific microorganisms and include protocols for individual microorganisms ranging from pathogens *Salmonella* to *Bifidobacterium*, a microorganism considered beneficial to humans and animals. In conjunction with the style of the Methods in Molecular Biology series, the respective chapters provide stepwise laboratory protocols, lists of the necessary materials and reagents, and tips on troubleshooting and avoiding known pitfalls. Figures are provided where appropriate to illustrate examples of what experimental outcomes would look like and to add more understanding of the protocol being described. The second section covers more general protocols including plasmid transfer and bioinformatic tools as well as novel applications of transposon methodologies such as transposon-aided capture of antibiotic-resistant plasmids. It is anticipated that the protocols described in this book will serve as springboard for further development of transposon methods for microorganisms.

This book, entitled *Microbial Transposon Mutagenesis: Protocols and Applications*, serves as a reference source for anyone wanting to apply transposon mutagenesis to microbial genetic analyses and functionality.

Fayetteville, AR, USA — *Steven C. Ricke*
Corvallis, OR, USA — *Si Hong Park*
Fayetteville, AR, USA — *Morgan L. Davis*

Contents

Part II Transposon General Protocols and Applications

Contributors

HOSSAM ABDELHAMED • *College of Veterinary Medicine, Mississippi State University, Mississippi State, MS, USA*

JOHN P. BANNANTINE • *USDA-ARS-National Animal Disease Center, Ames, IA, USA*

RAÚL G. BARLETTA • *School of Veterinary Medicine and Biomedical Sciences, University of Nebraska, Lincoln, NE, USA*

KODY A. BASSETT • *Department of Biological Sciences, Missouri University of Science and Technology, Rolla, MO, USA*

KENNETH W. BAYLES • *Department of Pathology and Microbiology, Center for Staphylococcal Research, University of Nebraska Medical Center, Omaha, NE, USA*

MICHAEL BLACK • *Biological Sciences Department, California Polytechnic University, San Luis Obispo, CA, USA*

DAMAYANTI CHAKRAVARTY • *Department of Cell and Molecular Biology, The University of Southern Mississippi, Hattiesburg, MS, USA*

REENA CHANDRASHEKAR • *Department of Poultry Science, University of Arkansas, Fayetteville, AR, USA*

SARAH DELANEY • *Antimicrobial Resistance and Microbiome Research Group, Department of Biology, Maynooth University, Maynooth, Co. Kildare, Ireland; Alltech, Dunboyne, Co. Meath, Ireland*

JANET R. DONALDSON • *Department of Cell and Molecular Biology, The University of Southern Mississippi, Hattiesburg, MS, USA*

JENNIFER L. ENDRES • *Department of Pathology and Microbiology, Center for Staphylococcal Research, University of Nebraska Medical Center, Omaha, NE, USA*

PAUL D. FEY • *Department of Pathology and Microbiology, Center for Staphylococcal Research, University of Nebraska Medical Center, Omaha, NE, USA*

PAT M. FIDOPIASTIS • *Biological Sciences Department, California Polytechnic University, San Luis Obispo, CA, USA*

STEVEN L. FOLEY • *U.S. Food and Drug Administration, National Center for Toxicological Research, Jefferson, AR, USA*

RONALD L. FRANK • *Department of Biological Sciences, Missouri University of Science and Technology, Rolla, MO, USA*

MICHAEL GIVSKOV • *Department of Immunology and Microbiology, Faculty of Health and Medical Sciences, Costerton Biofilm Center, University of Copenhagen, Copenhagen, Denmark; Singapore Center for Environmental Life Sciences Engineering, Nanyang Technological University, Singapore, Singapore*

FASAHATH HUSAIN • *Greater Los Angeles VA Health Care System, Los Angeles, CA, USA*

KWANGCHEOL C. JEONG • *Department of Animal Science, Institute of Food and Agricultural Sciences, University of Florida, Gainesville, FL, USA*

TIESHAN JIANG • *Department of Poultry Science, University of Arkansas, Fayetteville, AR, USA*

PRAVIN R. KALDHONE • *U.S. Food and Drug Administration, National Center for Toxicological Research, Jefferson, AR, USA*

SARDAR KARASH • *Cell and Molecular Biology Program, University of Arkansas, Fayetteville, AR, USA*

Attila Karsi • *College of Veterinary Medicine, Mississippi State University, Mississippi State, MS, USA*

Bijay K. Khajanchi • *U.S. Food and Drug Administration, National Center for Toxicological Research, Jefferson, AR, USA*

Young Min Kwon • *Cell and Molecular Biology Program, University of Arkansas, Fayetteville, AR, USA; Department of Poultry Science, University of Arkansas, Fayetteville, AR, USA*

Mark L. Lawrence • *College of Veterinary Medicine, Mississippi State University, Mississippi State, MS, USA*

Sang In Lee • *Department of Food Science and Technology, Oregon State University, Corvallis, OR, USA*

F. Chris Minion • *Department of Veterinary Microbiology and Preventive Medicine, Iowa State University, Ames, IA, USA*

Ian Moppert • *Department of Food Science and Technology, Oregon State University, Corvallis, OR, USA*

Melanie R. Mormile • *Department of Biological Sciences, Missouri University of Science and Technology, Rolla, MO, USA*

Richard Murphy • *Alltech, Dunboyne, Co. Meath, Ireland*

Martin Nilsson • *Department of Immunology and Microbiology, Faculty of Health and Medical Sciences, Costerton Biofilm Center, University of Copenhagen, Copenhagen, Denmark*

Si Hong Park • *Department of Food Science and Technology, Oregon State University, Corvallis, OR, USA*

Oindrila Paul • *Department of Cell and Molecular Biology, The University of Southern Mississippi, Hattiesburg, MS, USA*

Supraja Puttamreddy • *Merck, Ames, IA, USA*

Bryna Rackerby • *Department of Microbiology, Oregon State University, Corvallis, OR, USA*

Philip N. Rather • *Department of Microbiology and Immunology, Emory University, Atlanta, GA, USA; Research Service, Atlanta VA Medical Center, Decatur, GA, USA*

Steven C. Ricke • *Department of Food Science, Center for Food Safety, University of Arkansas, Fayetteville, AR, USA; Cell and Molecular Biology Program, University of Arkansas, Fayetteville, AR, USA*

Lorena Ruiz • *Department of Microbiology and Biochemistry of Dairy Products, IPLA-CSIC, Villaviciosa, Asturias, Spain*

Deepti Samarth • *Department of Poultry Science, University of Arkansas, Fayetteville, AR, USA*

Eric V. Stabb • *Department of Microbiology, University of Georgia, Athens, GA, USA*

Julie L. Stoudenmire • *Department of Microbiology, University of Georgia, Athens, GA, USA*

Tim Tolker-Nielsen • *Department of Immunology and Microbiology, Faculty of Health and Medical Sciences, Costerton Biofilm Center, University of Copenhagen, Copenhagen, Denmark*

Douwe van Sinderen • *APC Microbiome Ireland, University College Cork, Cork, Ireland; School of Microbiology, University College Cork, Cork, Ireland*

Yaligara Veeranagouda • *Greater Los Angeles VA Health Care System, Los Angeles, CA, USA; David Geffen School of Medicine, University of California Los Angeles, Los Angeles, CA, USA*

FIONA WALSH • *Antimicrobial Resistance and Microbiome Research Group, Department of Biology, Maynooth University, Maynooth, Co. Kildare, Ireland*
HANNAH M. WEXLER • *Greater Los Angeles VA Health Care System, Los Angeles, CA, USA; David Geffen School of Medicine, University of California Los Angeles, Los Angeles, CA, USA*
VIJAYA KUMAR YAJJALA • *Department of Pathology and Microbiology, Center for Staphylococcal Research, University of Nebraska Medical Center, Omaha, NE, USA*
YICHAO YANG • *Department of Poultry Science, University of Arkansas, Fayetteville, AR, USA*
DENISE K. ZINNIEL • *School of Veterinary Medicine and Biomedical Sciences, University of Nebraska, Lincoln, NE, USA*

Part I

Transposon Protocols for Specific Microorganisms

Chapter 1

Preparation of Transposon Library and Tn-Seq Amplicon Library for *Salmonella* Typhimurium

Sardar Karash, Tieshan Jiang, Deepti Samarth, Reena Chandrashekar, and Young Min Kwon

Abstract

Transposon sequencing (Tn-seq) is a powerful tool for functional genomics of bacteria. Tn-seq combines transposon mutagenesis with next generation sequencing to assess genetic requirements at a genome-wide scale and identify essential and conditionally essential genes. An efficient application of this experimental approach relies on robust protocols for transposon mutagenesis system and Tn-seq amplicon library preparation method. However, the existing approaches for the Tn-seq amplicon library preparation have several shortcomings. Hence, we present a robust, fast, specific, and cost-effective approach for the transposon mutagenesis of *Salmonella* Typhimurium and Tn-seq amplicon library preparation for Illumina sequencing. Besides *S.* Typhimurium that was used here for illustration, this protocol can also be used for other bacteria. In particular, the procedure for Tn-seq amplicon library preparation can be broadly applicable to any transposon elements. We delineate comprehensive step-by-step protocols for transposon mutagenesis and Tn-seq amplicon library such that it can be reproduced effortlessly by other researchers.

Key words Transposon mutagenesis, Tn-seq amplicon library, Illumina sequencing, Functional genomics, *Salmonella* Typhimurium

1 Introduction

With advances in DNA sequencing technologies, the whole genome sequence of a bacterium can be obtained routinely. However, we still need tools to direct these genotypes to their relevant phenotypes, including virulence and antibiotic resistance phenotypes as well as unknown molecular pathways. Coupling transposon mutagenesis with Illumina sequencing allows for a comprehensive phenotypic assessment of hundreds of thousands of mutants simultaneously. This method, Tn-seq, is a powerful tool to assess fitness requirement of each gene on the entire genome of a prokaryotic organism [1, 2]. In Tn-seq method, a saturated transposon insertion library (input pool) is exposed to a selective condition, and the mutant population altered through the selection (output pool) is

Steven C. Ricke et al. (eds.), *Microbial Transposon Mutagenesis: Protocols and Applications*, Methods in Molecular Biology, vol. 2016, https://doi.org/10.1007/978-1-4939-9570-7_1,

recovered. The genomic junctions of the transposon insertions are then specifically amplified and sequenced from both input and output pools, by a high-throughput sequencing technology. The mutant fitness for each gene can be calculated by the change in relative abundance of the sequence reads corresponding to the insertion mutants during the selection. Thus, Tn-seq can identify the majority, if not all, of the genes in a bacterial genome that are specifically required for fitness under a given condition of interest (e.g., the genes required for resistance against host stressors in the case of bacterial pathogens).

Recently we used a mini-transposon delivery vector pBAM1 for transposon mutagenesis of *S.* Typhimurium [3], which was originally developed for mutagenesis of *Pseudomonas putida* [4]. We then developed a new method for Tn-seq amplicon library preparation that has several advantages over existing methods. First, it requires only approximately 100 ng of the genomic DNA extracted from a mutant library. Second, we developed a facile method to eliminate the sequence reads from pseudo mutants produced by integration of the entire plasmid pBAM1 into the chromosome. These pseudo mutants would consume a significant number of Tn-seq sequence reads, which consequently reduce the resolution of the overall Tn-seq analysis. Third, to increase the specificity in amplifying the Tn-seq genomic junctions in the library, we employed a dual priming oligonucleotide (DPO) [5].

Particularly, the protocol for Tn-seq amplicon library preparation has broad applicability to any transposon elements in any bacterial species. We have successfully used this protocol for the libraries constructed using various transposon elements in the background of *S.* Typhimurium [3, 6, 7], and *Campylobacter jejuni* [8].

2 Materials

2.1 Bacterial Strains and Plasmids

1. *Escherichia coli* SM10 λ*pir* carrying a transposon-delivery plasmid vector pBAM1 (Amp^R and Km^R) as a donor strain (*see* **Note 1**).
2. *Salmonella enterica* serovar Typhimurium 14028s (with spontaneous mutation conferring resistance to NA) as a recipient strain (*see* **Note 2**).

2.2 Antibiotics, Buffers, Solutions, Media, and Filters (See Note 3)

1. 25 μg/ml nalidixic acid (NA).
2. 50 μg/ml ampicillin (Amp).
3. 50 μg/ml kanamycin (Km).
4. 10 mM $MgSO_4$.
5. Phosphate-buffered saline (PBS), pH 7.4.
6. Tris–EDTA buffer (pH 8.0).

7. Luria–Bertani (LB) broth.
8. LB agar.
9. Dimethyl sulfoxide (DMSO).
10. 0.45-μm 47 mm Nitrocellulose analytical test funnel filters (Nalgene™, Thermo Scientific).
11. Elution buffer (EB; 10 mM Tris–HCl, pH 8.5).
12. 2.5 mM $CoCl_2$.
13. 10 mM dCTP.
14. 1 mM ddCTP.

2.3 Kits

1. DNeasy Blood & Tissue kit (Qiagen, Valencia, CA).
2. Qubit™ dsDNA BR Assay kit (Invitrogen™).
3. DNA Clean & Concentrator-5 kit (Zymo Research).
4. GoTaq Colorless Master Mix (Promega).
5. Zymoclean Gel DNA Recovery kit (Zymo Research).

2.4 Enzymes

1. Terminal transferase (TdT; New England Biolabs).
2. PvuII (New England Biolabs (*see* **Note 4**).

3 Methods

The protocols presented in this chapter consist of two parts: (1) Preparation of mutant pools for Tn-seq analysis, and (2) Preparation of Tn-seq amplicon library. In the first part, a library of transposon mutant is constructed (input pool), which then is subjected to selection under the selective conditions of interest (output pools). The details of the construction of the library largely depend on the method of choice for transposon mutagenesis established for the bacterial species under the study. In this chapter, we provided the protocol for Tn5 mutagenesis we have used in our recent studies [3, 7] using Tn5 delivery vector pBAM1 via conjugation [4]. However, researchers can choose different transposon mutagenesis system suitable for the bacterial species of interest. The experimental design and details of the mutant selection would also depend on the research questions and focus of the researches. In the second part, the mutants pools thus prepared are subjected to Tn-seq amplicon library protocol. Although there are multiple approaches for this, we are presenting a versatile protocol that we have developed and used in our recent studies [3, 6–8]. Our Tn-seq amplicon library protocol illustrated in Fig. 1 can be universally applied to any transposon elements with appropriate changes in the design of the PCR primers used in the protocol.

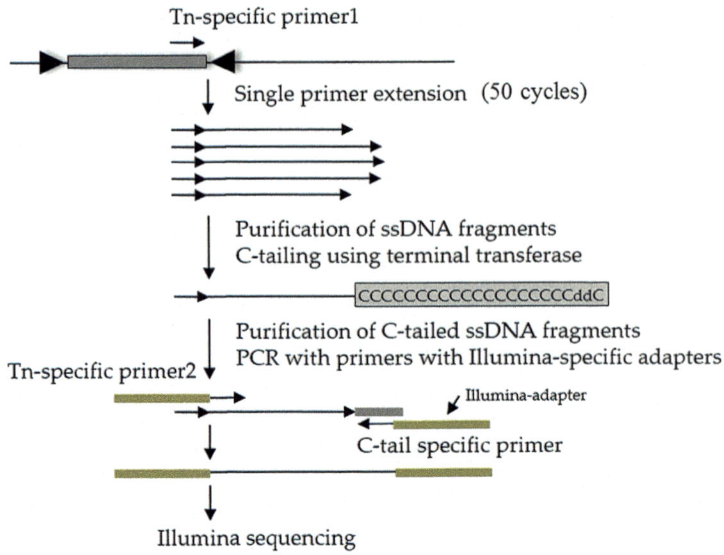

Fig. 1 The strategy for preparation of Tn-seq amplicon library. The genomic DNA is extracted from the mutant pool, and then subjected to the protocol as illustrated here. First step is a linear extension using a single primer specific to one end of Tn5 to capture Tn5 junctions. The second step is to add C-tail at 3′ end of the captured single-stranded Tn5 junction fragments. Lastly, the C-tailed Tn5 junction fragments are used as a template for exponential PCR using nested Tn5-specific primer and poly G primer with attached Illumina adaptors and a sample index. The PCR products are then gel-purified and sequenced by Illumina HiSeq

3.1 Transposon Mutagenesis via Biparental Mating

Always use fresh bacterial cultures for streaking and subculturing (*see* **Note 5**).

3.1.1 Day One

1. Inoculate a single colony of the donor strain, *E. coli* SM10 λ*pir*, that harbors the plasmid pBAM1 in LB broth supplemented with Amp (50 μg/ml) and Km (50 μg/ml), and the recipient bacterial strain *S.* Typhimurium 14028s (NA^R) in LB broth supplemented with NA (25 μg/ml).
2. Incubate at 250 rpm, 37°C overnight (~18 h).

3.1.2 Day Two

1. Add 1 ml of donor strain and recipient bacterial strain in separate 2 ml sterile microcentrifuge tubes from overnight grown cultures. Centrifuge at 8228 × *g* for 1 min and resuspend in 1 ml LB broth without antibiotics (*see* **Note 6**).
2. Combine equal volumes (1 ml) of the donor and recipient strain in a 5 ml test tube with a snap cap. Pipet up and down 4 to 5 times or vortex to mix.

3. Pellet the 2 ml bacterial mixture at 8228 × *g* for 1 min. Discharge the supernatant using a pipette.
4. Resuspend the pellet in 2 ml of 10 mM $MgSO_4$ by vortexing. Spin at 8228 × *g* for 1 min. Remove the supernatant using a pipette.
5. Resuspend the pellet in 2 ml PBS (*see* **Note 7**). Vortex to mix.
6. Using gentle vacuum pressure, filter the resuspended cell mixture (2 ml) through a funnel filter containing a 0.45-μm pore 47 mm filter (*see* **Note 8**). The purpose of this step is to spread the cell mixture evenly on the surface of the filter disc where conjugation can happen.
7. Using a sterile forceps, carefully remove the filter from the funnel filter assembly and place it on a LB agar plate (no antibiotic) with the bacterial cells facing up.
8. Incubate for 3–5 h at 37 °C (*see* **Note 9**).
9. Place the filter in a 15 ml Falcon tube containing 5 ml of 10 mM $MgSO_4$.
10. To release the bacterial cells, vortex for 1 min.
11. Transfer the cell suspension in 10 mM $MgSO_4$ solution to a new 15 ml Falcon tube.
12. Concentrate the cell suspension by spinning the supernatant at 8228 × *g* for 2 min. Remove the supernatant using a pipette, and resuspend the pellet in 1 ml of 10 mM $MgSO_4$.
13. Pipet 100 μl of the conjugation mixture per LB agar plate supplemented with 50 μg/ml Km and 25 μg/ml NA (*see* **Note 10**).
14. Spread the bacterial cells evenly on the agar plates using a sterile L-shaped spreader. Alternatively, dispense 10–15 sterile 4 mm glass plating beads per plate and shake the plates back and forth a few times to ensure even spreading of the bacterial cells. Remove the glass beads to be used again later.
15. Incubate the plates 37 °C for approximately 24 h.

3.1.3 Day Three

1. To harvest the mutants, add 1 ml LB broth containing 50 μg/ml Km and 7% DMSO to each plate.
2. Scrape the colonies with a sterile L-shaped spreader and collect the cells using a pipette.
3. Combine all mutant cultures from all plates into a 50 ml Falcon tube. Mix well the mutants by vortexing (*see* **Note 11**).
4. Aliquot 500 μl of the transposon mutant library into cryogenic tubes and store them at −80 °C (*see* **Note 12**).

3.2 Checking the Frequency of Pseudo Mutants

1. Prepare LB agar plates containing Amp (50 μg/m).
2. Divide the back of the plates to 50 squares with a marker.
3. Using sterile toothpicks pick 100 random colonies from the plates of Subheading 3.2 (Day 2 No. 15) and streak onto the squares of LB agar plates (Amp).
4. Incubate at 37 °C for approximately 18 h.
5. Count the number of squares with mutant growth (*see* **Note 13**).
6. Calculate the percentage of pseudo mutants according to the number of colonies grown on the LB agar plates (Amp).

3.3 Selection of Transposon Mutant Library

For Tn-seq experiments, the transposon mutant library in the stock stored at −80 °C would be often used. To use such a library, appropriate steps for "activation" of the library to make the mutants ready for the selection of the study are necessary.

3.3.1 Washing and Activation of the Mutant Library

1. Thaw a transposon library stock culture on ice.
2. Dilute 500 μl of the thawed stock culture ten times by adding it into fresh 4.5 ml warm LB broth supplemented with Km (50 μg/ml).
3. Incubate at 37 °C for 1 h with vigorous shaking at 220 rpm.
4. Spin at 1157 × *g* for 8 min.
5. Remove the supernatant.
6. Resuspend the pellet in 5 ml PBS.
7. Spin at 1157 × *g* for 8 min.
8. Repeat the washing **steps 5**–7.
9. Resuspend the pellet in 5 ml LB medium.
10. Check the number of cells in this suspension by plating serial dilutions on LB agar plates supplemented with Km (50 μg/ml). After washing and activation the yield should be approximately 2.0×10^{11} CFUs/ml (*see* **Note 14**).

3.3.2 Selection of the Mutant Library

The details of the experimental design for the selection would be largely dependent on the goal of the study. However, there are two main aspects that are important for accurate identification of the genes by minimizing the chance for false positive or false negative gene discoveries. First, the inoculum should contain the number of bacterial cells that would represent each mutant in the library by multiple cells. We typically demand 10 cells/mutant for the calculation of the CFUs in the inoculum. Second, the experiment should be designed to eliminate the chance for mutants to disappear during selection by random chance but not due to the mutant phenotype. This issue of "bottleneck" is often problematic when animal models are used as selection medium for Tn-seq screening of bacterial pathogens. For more information on this bottleneck issue related to in vivo selection, the readers are advised to refer to a review article by Abel et al. [9]

3.4 Preparation of Tn-Seq Amplicon Library for Illumina Sequencing

Our protocol for the preparation on Tn-seq amplicon library is illustrated in Fig. 1. The first step is a linear extension of Tn-junction sequences from gDNA of the mutant pool using a primer specific to one end of the transposon (Tn-specific primer1 in Fig. 1; Tn5-DPO in Table 2). Then, C-tail is attached to 3′ end of each extended single-stranded DNA fragment, which provides the annealing site for C-tail specific primer (P7-16G primer in Table 2). Now single-stranded C-tailed Tn-junction fragments are efficiently amplified via PCR reaction using a nested transposon-specific primer (Tn-specific primer2 in Fig. 1; P5-BC1-Tn5-O in Table 2) and C-tail specific primer (P7-16G primer in Table 2).

3.4.1 Restriction Enzyme Digestion to Remove Reads from Pseudo Mutants (Optional)

PvuII enzyme sites are present immediately outside the inverted repeats on both sides of but not within the Tn5 element in pBAM1 plasmid. Therefore, the genomic DNA can be digested with this enzyme to effectively eliminate Tn-junction sequences originated from pseudo mutants (*see* **Note 4**).

1. Digest approximately 300 ng of the gDNA using PvuII overnight on a thermal cycler or incubator, following the manufacturer's instructions for restriction enzyme digestion.
2. Purify the digested DNA by using DNA Clean & Concentrator-5 kit.
3. Proceed to the next step or store at −20 °C.

3.4.2 Linear Extension Using a Transposon-Specific Primer

1. Perform a linear extension PCR following Table 1 using gDNA from input and output pools. Also, include gDNA of the wild-type strain as a negative control.
2. The use of DPO primer is recommended to increase the specificity in amplification of Tn-seq genomic junction sequence (*see* **Note 15**). Table 2 gives the primer sequences.
3. Perform the PCR cycle as shown in Table 3 (*see* **Note 16**).

Table 1
Linear extension PCR reaction mixture

GoTaq Green Master Mix	25.0 μl
Tn5-DPO (20 μM)	1.0 μl
Genomic DNA of Tn library (50 ng/μl)	2.0 μl
ddH_2O	22.0 μl
Total	50.0 μl

Table 2
Oligonucleotides

Tn5-DPO Tn-specific primer1)	AAGCTTGCATGCCTGCAGGTIIIIICTAGAGGATC
P5-BC1-Tn5-O[a] (Tn-specific primer2)	**AATGATACGGCGACCACCGAGATCTACACTCTTTCCCTAC** **ACGACGCTCTTCCGATCT**NNNNAGATCACGCCTAGGCGGC CTTAATTAAAGATGTGTATAAGAG
P7-16G[a] (C-tail-specific primer)	**CAAGCAGAAGACGGCATACGAGCTCTTCCGATCT**GGGGGGGGGGGGGGGG

[a]P5 and P7 sequences are indicated in bold. Barcode is indicated by underline. Sample index (or barcode, BC) "ATCAC" is shown for P5-BC1-Tn5-O, but different sample indices are used for different samples.

Table 3
PCR cycles for linear extension

	95 °C	2 min
	95 °C	30 s
50 Cycles	62 °C	45 s
	72 °C	10 s
	4 °C	Hold

4. Purify the linear extension PCR products (single-stranded DNA) using DNA Clean & Concentrator-5 kit. Elute DNA in 15 μl EB buffer. Proceed to the next step or store the eluted DNA at −20 °C.

3.4.3 C-Tailing Reaction

At this step we add C-tails to the purified linear PCR products using terminal transferase (TdT) (Fig. 1). The C-tails serve as the annealing site for poly G primer to allow for exponential amplification of Tn-junction sequences.

1. Prepare the reaction mixture following Table 4.
2. Incubate the reaction at 37 °C for 1 h.
3. Heat-inactivate TdT enzyme by incubation at 75 °C 10 min.
4. Purify the C-tailed products using DNA Clean & Concentrator-5 kit. Elute DNA in 10 μl EB buffer. Proceed to next step or store C-tailed DNA at −20 °C.

3.4.4 Exponential Amplification of Tn-Flanking Sequences

This step is to amplify the purified C-tailed products with an exponential PCR reaction using two primers to produce PCR products compatible for Illumina sequencing (Fig. 1). To allow for multiplexing, sample index sequences are also included in the design of the oligonucleotides. Figure 2 shows an example of

Table 4
C-tailing reaction mixture

TdT Buffer (10×)	2.0 μl
2.5 mM $CoCl_2$	2.0 μl
10 mM dCTP	2.4 μl
1 mM ddCTP	1.0 μl
Terminal transferase (TdT)	0.5 μl
DNA (linear extension products)	13.0 μl
Total	20.0 μl

AATGATACGGCGACCACCGAGATCTACACTCTTTCCCTACACGACGCTCTTCCGA
TCTNNNNAGATCACGCCTAGGCGGCCTTAATTAAAGATGTGTATAAGAGACAGGC
GCCATAACGCGCTTACTTTTACGAACTGAAAGCGAAGAAGGAGGAAACAATTCAG
GGCATTCACTTAGCCAGGAGTTCCACAACATGTATCCCAGCGTGGTATGCGCGAT
ATCCCCCCCCCCCCCCCCCCC**AGATCGGAAGAGCTCGTATGCCGTCTTCTGCTTG**

Fig. 2 The structure and components of Tn-seq amplicon fragments. An example of Tn-seq amplicon fragment is shown to illustrate the important components. The two primers used for exponential amplification step, P5-BC1-Tn5-O and P7-16G primer, are highlighted in gray color. Inverted repeat sequence of Tn5 (Mosaic end; AGATGTGTATAAGAGACAG) is shown in red color. In P5-BA1-Tn5-O primer, random 4 nt region for efficient cluster identification on Illumina flowcells (NNNN) and a sample index sequence for demultiplexing (ATCAC) are underlined. The C-tail to which P7-16G primer anneals is also indicated in red color. Illumina adapter sequences P5 and P7 are highlighted in bold type. Tn5-junction sequence, which is used for genome-mapping to identify the insertion site, is highlighted in yellow color

Table 5
Exponential PCR reaction mixture

GoTaq Green master mix	25.0 μl
P5-BC1-Tn5-O (10 μM)	1.0 μl
P7-16G (10 μM)	1.0 μl
C-tailed DNA	1.0 μl
ddH_2O	22.0 μl
Total	50.0 μl

Sanger DNA sequencing result from a cloned fragment in the amplicon library.

1. Follow Table 5 for PCR reaction mixture.
2. Follow the Exponential PCR program in Table 6.
3. Proceed to next step or store samples at −20 °C.

Table 6
PCR cycles for exponential PCR

	95 °C	2 min
	95 °C	30 s
36 Cycles	58 °C	45 s
	72 °C	20 s
	72 °C	10 min
	4 °C	Hold

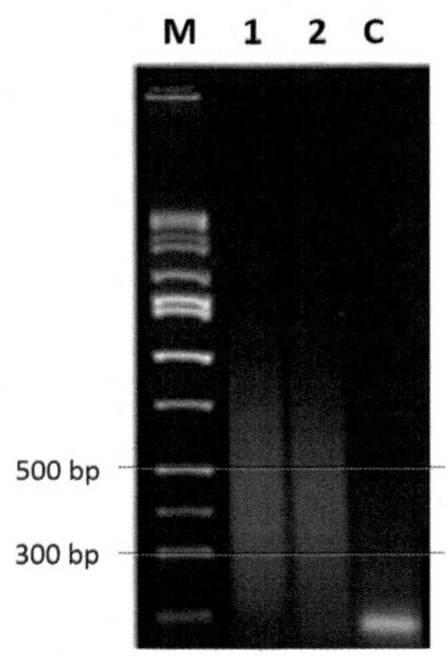

Fig. 3 Size-selection of Tn-seq amplicon libraries. After the final step of the Tn-seq amplicon library preparation, the PCR products are separated on a 1.5% agarose gel. The gel pieces containing DNA fragments ranging between 300 and 600 bp are excised and used to extract DNA fragments, which are then used for Illumina sequencing: M. Hi-Lo Marker, 1. Mutant pool 1, 2. Mutant pool 2, and C. Negative control (the wild-type strain)

3.4.5 Agarose Gel Purification

1. Incubate the 50 μl PCR products at 60 °C for 15 min and place directly on ice for 5 min (*see* **Note 17**).
2. Load the 50 μl of exponential PCR products on 1.5% agarose gel in 0.5% TAE buffer.
3. Run the gel for 60 min at 100 V.
4. View the gel on a UV transilluminator and excise the target fragment size of 300–500 bp (Fig. 3). We should expect to see smear patterns of DNA fragments. The negative control using the wild-type strain should not show any background amplification as shown in Fig. 3.

5. Gel-purify DNA fragments using Zymoclean Gel DNA Recovery kit.
6. Samples are ready for sequencing on Illumina next generation sequencing platforms (*see* **Note 18**).
7. Store the samples at −20 °C.

4 Notes

1. This Tn5 delivery system may be broadly applicable to many other Gram negative bacterial species for efficient Tn5 mutagenesis.
2. We used the nalidixic acid resistance (NA^R) as a counterselection marker to select the recipient strain against the donor strain. Alternative antibiotic resistance can be used for this purpose.
3. Indicated are the final working concentrations.
4. The restriction enzyme PvuII is specific to the delivery plasmid pBAM1 described in this chapter for illustration. This restriction enzyme cuts immediately outside both ME-O and ME-I regions in pBAM1 [4]. Suitable restriction enzyme may or may not be available depending on the specific transposon delivery vectors to be used.
5. Bacteria strains are streaked on agar plates from a stock cultures stored −80 °C. Streaked plates can be stored at 4 °C for approximately 2 weeks.
6. This step is necessary to remove antibiotics from the cultures of both donor and recipient strains.
7. The pellet can also be resuspended in 10 mM $MgSO_4$ but a higher yield has been observed with PBS.
8. Using vacuum pressure for a long time can dehydrate bacterial cells and may cause damage to filter. Also, in the case of unavailability of funnel filter assembly, any cellulose nitrate filter discs with the pore size ≤0.45 μm can be used for this purpose. Place the cellulose filter disc on top of sterile blotting paper or Whatman™ filter paper. Spread the bacteria cells in PBS gently over the filter disc leaving the perimeter of filter disc dry. Wait until excess PBS is absorbed by the blotting paper leaving minimal moisture over the filter.
9. The incubation time for mutagenesis via conjugation can vary. Longer incubation leads to a high number of mutants but also increases the number of undesirable siblings carrying the same insertions.

10. To enumerate the number of mutants, the cell suspension needs to be serially diluted and plated on LB agar plates supplemented with Km and NA. To maximize the number of mutants recovered, this conjugation mix can be plated on to multiple LB agar plates supplemented with Km and NA.
11. Typically, 1 ml reaction of donor and recipient strain yields about 70,000 mutants based on plate counting enumeration in our experience. To increase the saturation level of the library, multiple conjugation reactions can be performed in parallel. However, all these mutants that are resistant to Km and NA are not true Tn5 mutants but can also include pseudo mutants.
12. We used DMSO at a final concentration of 5–15% (v/v) as a cryoprotectant because it is less viscous than glycerol, making it easier to handle. However, glycerol is less toxic to cells than DMSO.
13. The pseudo mutants in which the entire plasmid pBAM1 is integrated into the chromosome are resistant to Amp in addition to Km, and NA.
14. It is important to know the CFUs of the starter library, as it reflects the cell density of the mutant library. This information enables estimation for the number of mutants that would be included in mutant selection, and number of sequencing reads that need to be obtained from each mutant pool.
15. We found the use of Tn5-specific primer designed according to the principle of DPO primer [5] significantly increased the percentage of genome-mapping (93.1%) as compared to the normal primer design (75.4%), indicating significantly reduced nonspecific amplification [8].
16. We used hot-start PCR by manually adding *Taq* DNA polymerase during the first denaturation step to eliminate nonspecific amplification. This can be done using alternative hot-start protocols.
17. This heating step is optional but is known to improve better separation of the fragments according to the lengths.
18. We prefer to sequence the Tn-seq library in HiSeq platforms at 151 cycles because this provides a high number of reads, which would be critical for accurate quantitative profiling of transposon mutants. Additionally, the length of the genomic junction region was greater than 90 nucleotides in our experimental design when sequenced through 151 cycles, which allowed for accurate genome mapping of Tn junction sequence reads.

References

1. van Opijnen T, Bodi KL, Camilli A (2009) Tn-seq: high-throughput parallel sequencing for fitness and genetic interaction studies in microorganisms. Nat Methods 6:767–772. https://doi.org/10.1038/nmeth.1377
2. Kwon YM, Ricke SC, Mandal RK (2016) Transposon sequencing: methods and expanding applications. Appl Microbiol Biotechnol 100 (1):31–43. https://doi.org/10.1007/s00253-015-7037-8
3. Karash S, Liyanage R, Qassab A, Lay JO, Kwon YM (2017) A comprehensive assessment of the genetic determinants in *Salmonella* Typhimurium for resistance to hydrogen peroxide using proteogenomics. Sci Rep 7:17073. https://doi.org/10.1038/s41598-017-17149-9
4. Martínez-García E, Calles B, Arévalo-Rodríguez M, de Lorenzo V (2011) pBAM1: an all-synthetic genetic tool for analysis and construction of complex bacterial phenotypes. BMC Microbiol 11(1):38. https://doi.org/10.1186/1471-2180-11-38
5. Chun JY, Kim KJ, Hwang IT, Kim YJ, Lee DH, Lee IK, Kim JK (2007) Dual priming oligonucleotide system for the multiplex detection of respiratory viruses and SNP genotyping of CYP2C19 gene. Nucleic Acids Res 35:e40. https://doi.org/10.1093/nar/gkm051
6. Mandal RK, Kwon YM (2017) Global screening of *Salmonella enterica* serovar Typhimurium genes for desiccation survival. Front Microbiol 8:1723. https://doi.org/10.3389/fmicb.2017.01723
7. Karash S, Kwon YM (2018) Iron-dependent essential genes in *Salmonella* Typhimurium. bioRxiv:159921. https://doi.org/10.1101/159921
8. Mandal RK, Jiang T, Kwon YM (2017) Essential genome of *Campylobacter jejuni*. BMC Genomics 18:616. https://doi.org/10.1186/s12864-017-4032-8
9. Abel S, Abel zur Wiesch P, Davis BM, Waldor MK (2015) Analysis of bottlenecks in experimental models of infection. PLoS Pathog 11:e1004823. https://doi.org/10.1371/journal.ppat.1004823

Chapter 2

Application of Whole-Genome Sequencing to Transposon Mutants of *Salmonella* Heidelberg

Bryna Rackerby, Sang In Lee, Ian Moppert, Steven C. Ricke, KwangCheol C. Jeong, and Si Hong Park

Abstract

Transposons are elements widely dispersed among organisms which are able to move and replicate fragments of genomes. The extensive variability in transposons present in most organisms requires extensive identification and interpretation of the resulting transposon mutants after transposon mutagenesis. However, much of this is reliant on utilizing randomness characteristics of transposon to identify essential genes for the organism of interest. Integration of the transposon mutant approach with commercialized next-generation sequencing (NGS) technology has helped to advance transposon identification by sequencing millions of reads generated from a single run on an NGS platform. Transposon sequencing is defined as a gene sequencing methodology that allows for the identification of nonessential genes and the determination of gene function using a random transposon insertional mutagenesis followed by massively parallel sequencing. The detailed protocol will be outlined in this chapter. The genomic DNA integrated with the transposons is sequenced using an NGS platform in order to determine the location of each mutation.

Key words *Salmonella* Heidelberg, Transposon mutant, Next-generation sequencing (NGS), Whole-genome sequencing (WGS)

1 Introduction

Transposons were first identified by a geneticist named Barbara McClintock more than 50 years ago in New York [1]. Transposons were commonly known as "jumping genes" due to their ability to relocate their sequences on the genomic or plasmid DNA. They can be categorized into two classes (1 and 2) and subcategorized as autonomous or nonautonomous transposons by their ability to move on their own [2]. Class 1 transposon, known as retrotransposon, is characterized by its mechanism of "copy and paste" and ability to produce RNA transcripts which rely on reverse transcriptase to be transcribed back into DNA [2, 3]. Class 2 transposon,

Bryna Rackerby and Sang In Lee are co-first authors and contributed equally to this work.

Steven C. Ricke et al. (eds.), *Microbial Transposon Mutagenesis: Protocols and Applications*, Methods in Molecular Biology, vol. 2016, https://doi.org/10.1007/978-1-4939-9570-7_2, © Springer Science+Business Media, LLC, part of Springer Nature 2019

known as DNA transposon is characterized by its ability of a "cut and paste" mechanism and encodes protein transposase which inserts and excises target fragment of gene without using RNA as an intermediate step [2, 3].

Transposons can be introduced into the organism of interest in multiple ways; transduction, introduction of DNA into a cell by a viral vector; transformation, direct uptake and incorporation through the cell membrane of DNA from the environment; transfection, introduction of naked nucleic acids by artificial means (into eukaryotes only; historically refers to uptake of viral DNA) [4]; and physical injection with a gene gun [5]. Biological methods of introducing foreign DNA into an organism tend to be more efficient and less damaging on DNA, although there are many physical methods of inserting the DNA such as electroporation, agitation with glass beads, ultrasound, and laser microbeams [6]. An accepted method of introducing carrier plasmids to the organism of interest is via conjugation by a donor organism carrying the plasmid, usually a strain of *Escherichia coli*. The conjugation method is less expensive and time-consuming, and requires less specialized equipment than running repeated rounds of electroporation [7].

During the past decade, development of next-generation sequencing (NGS) approaches by massive parallel sequencing technology exhibited exceptional ability among other sequencing methods in terms of read length, rate of read and quality of sequences. Besides the amplicon sequencing by NGS, one of the major sequencing methods integrated with NGS is whole-genome sequencing (WGS). WGS methods often employ an approach involving shotgun sequencing which is a metagenomics technique that assembles large number of various sized fragments generated by NGS platform to produce long reads of whole genomes. Since the development of NGS, Sanger sequencing method has essentially been replaced by NGS platforms for higher coverage and more reads in shorter periods of time [8].

The advent of NGS has advanced the sequencing of genomes at such a rate that the assessment of gene function cannot keep up, necessitating the creation of high throughput methods to more rapidly connect an already discovered genotype to its phenotype [9]. The solution to this problem is transposon sequencing, which utilizes transposon insertional mutagenesis and subsequent sequencing of saturated mutant libraries via high-throughput sequencing [10, 11] to link genes to their functions. Transposon insertional mutagenesis is where transposons are inserted into a target genome via physical, chemical, or biological methods. These insertions can be random, or can target specific locations in the genome. There are several methods for creating and analyzing saturated mutant libraries. The most common are Tn-Seq, INSeq, HITS, and TraDIS [9, 10]. The basic principle of all four methods involve purifying the genomic DNA, physical or enzymatic DNA fragmentation, attachment of adaptors for PCR amplification, and MPS (massively parallel sequencing) of the fragments [9, 10]. When physical shearing is used, the result is fragments of various sizes, potentially resulting in PCR bias toward amplification of smaller fragments [9].

2 Materials

2.1 Transposon Mutant Strain Construction

Salmonella Heidelberg isolates that are confirmed kanamycin sensitive strain (SH), EZ-Tn5™ pMOD™-6 <KAN-2/MCS> Transposon Construction Vector, (Epicentre), restriction enzyme (NEB, *Pvu*II-HF, blunt, CAGCTG), QIAquick gel extraction kit (Qiagen), electroporator cuvettes (0.4 cm), cold sterile 10% glycerol (keep always in a refrigerator), SOC media (Invitrogen), LB media (Difco Laboratories), LE agarose (Lonza), 1 kb DNA ladder (NEB), sterile razor blade, kanamycin (Thermo Fisher Scientific), ethidium bromide (EtBr, Sigma-Aldrich), UV transilluminator (Bio-Rad Laboratories), agarose gel electrophoresis system (Mupid), electroporator, incubator (VWR), water bath (VWR), NanoDrop (Thermo Fisher Scientific), spectrophotometer (Unico), centrifuge with refrigerated system (Eppendorf), vortexer (Scientific Industries).

2.2 Whole-Genome Sequencing (WGS)

DNeasy Blood and Tissue Kit (Qiagen) (*see* **Note 1**), Qubit™ dsDNA BR Assay Kit (Thermo Fisher Scientific), Nextera® XT DNA Library Prep Kit (Illumina) (*see* **Note 2**), MiSeq Reagent Kits v2 (2 × 250, 500 cycles, Illumina), PhiX V3 (Illumina), Qubit™ Assay Tubes (Thermo Fisher Scientific), 0.2 mL PCR tube (VWR), 10, 200 and 1000 μL multichannel pipettes (VWR), 96-well storage plates (Fisher Scientific), Agencourt AMPure XP (Beckman Coulter Genomics), DNase-RNase free water (VWR), Ethanol 200 proof (Sigma-Aldrich), 1N NaOH (VWR), reagent reservoirs (VWR), 96-well PCR plates (Eppendorf), 96-well seal film (VWR), Qubit Fluorometer (Thermo Fisher Scientific), heat block for 1.5 mL tubes (VWR), high-speed microplate shaker (VWR), magnetic stand-96 (Ambion), microcentrifuge (Eppendorf), vortexer (Scientific Industries), thermocycler (Eppendorf).

3 Methods

3.1 Transposon Mutant Construction

The described transposon mutant construction protocol is modified from a paper published by Khatiwara et al. [12].

3.1.1 Isolation Transposon Fragment from a Plasmid DNA

1. Restriction enzyme treatment to isolate transposon fragment from a plasmid DNA (EZ-Tn5™ pMOD™-6 <KAN-2/MCS> vector, Fig. 1) (*see* **Note 3**).

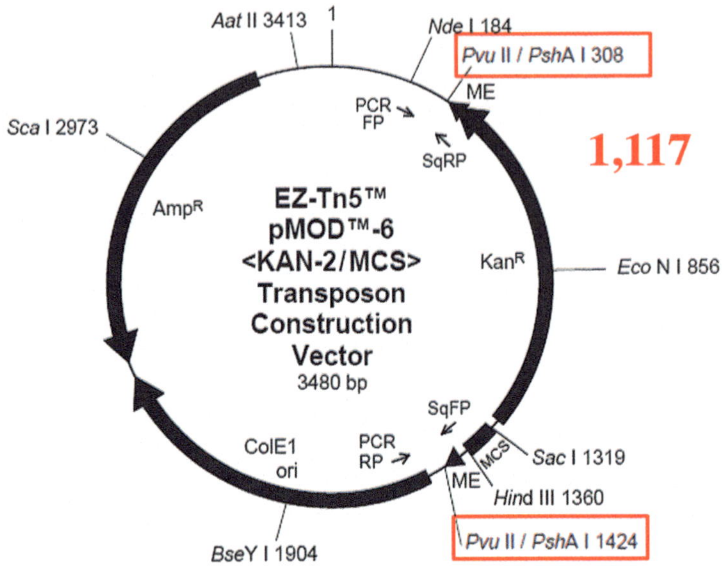

Fig. 1 Plasmid DNA (pMOD™-6) map included transposon fragment

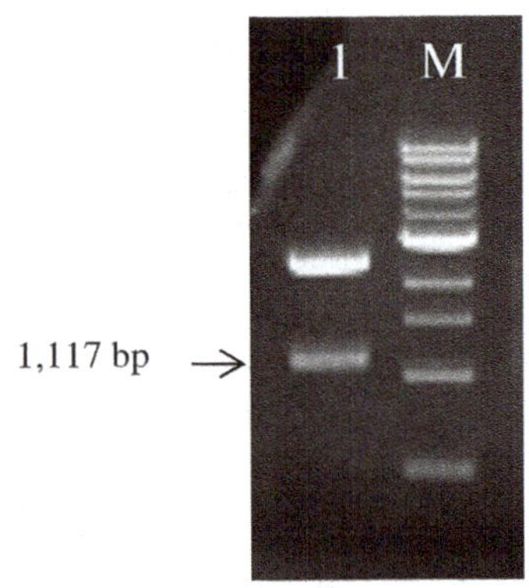

Fig. 2 Agarose gel electrophoresis after restriction enzyme treatment. Lane M: 1 kb DNA ladder, lane 1: pMOD™-6 plasmid DNA after restriction enzyme treatment

2. Mix 4 μg plasmid DNA (EZ-Tn5™ pMOD™-6 <KAN-2/MCS> vector), *Pvu*II restriction enzyme (10 U), 10 × buffer and DNase–RNase-free water up to 100 μL.
3. Incubate a mixture at 37 °C for 16 h.
4. Electrophorese restriction enzyme-treated mixture (10 μL) on the 1% agarose gel to confirm transposon fragment digestion (Fig. 2).
5. After **step 4**, make 1% agarose gel (narrow and wide wells) without an ethidium bromide (EtBr) (*see* **Note 4**).

Fig. 3 Agarose gel for recovery of transposon fragment. Left side of gel represents a 1 kb DNA ladder and digested plasmid DNA after EtBr staining. Right side of gel represents a cut transposon fragment without EtBr

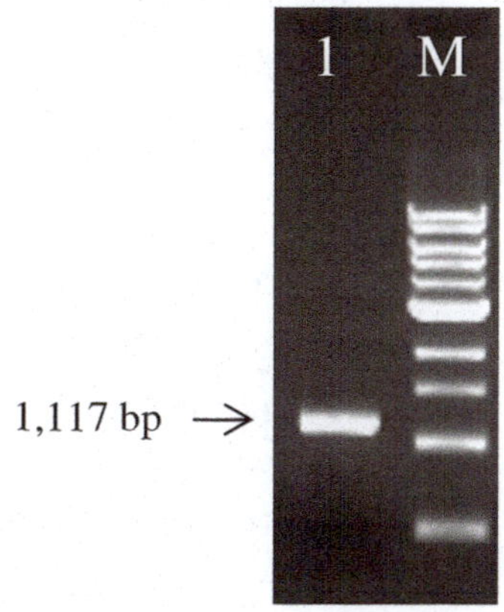

Fig. 4 Agarose gel after gel elution for confirmation. Lane M: 1 kb DNA ladder, lane 1: transposon fragment recovered from agarose gel (Fig. 3)

6. Electrophorese restriction enzyme-treated mixture on a narrow well (10 μL) and a wide well (80 μL), respectively.
7. Cut the agarose gel including the 1 kb DNA ladder and narrow well.
8. Stain the agarose gel portion from **step 7** with the EtBr solution for 30 min and compare with a wide well gel to locate transposon fragment on the UV transilluminator.
9. Cut the transposon fragment in the wide well of agarose gel using a razor blade (Fig. 3).
10. Gel extraction is carried out using a QIAquick Gel Extraction kit according to the manufacturer's instruction.
11. Confirm eluted transposon fragment (2 μL) from **step 10** on the 1% agarose gel (Fig. 4).
12. Keep transposon fragment at −20 °C until use.

3.1.2 Bacterial Growth and Make Competent Cells

1. Streak *Salmonella* Heidelberg that are confirmed kanamycin sensitive strain (SH) on a Luria–Bertani (LB) agar plate and incubate at 37 °C for 18 h.
2. Pick one colony, inoculate into 5 mL of LB broth, and incubate at 37 °C for 18 h.
3. Prepare LB agar plates supplemented with kanamycin (100 μg/ mL).
4. Make SH competent cells for transposon fragment transformation acquired from Subheading 3.1.1.
5. Inoculate 100 μL of overnight grown SH strain into 12 tubes of 10 mL LB media (no antibiotics) and incubate at 37 °C for 4 h.
6. Put the tubes including 4 h grown cells and sterile 10% glycerol in ice (*see* **Note 5**).
7. Centrifuge the culture tubes at 6800 × *g* for 10 min at 4 °C and discard the supernatant (combine 4 tubes to one 50 mL tube to make one competent cells).
8. Washing: add 5 mL sterile 10% glycerol to cells and swirl gently on ice to resuspend cells completely.
9. Repeat **steps** 7 and **8** for five times.
10. In the last washing step, add 1 mL sterile 10% glycerol and put the resuspension to a precold-adapted 1.5 mL tube.
11. Centrifuge (10,000 × *g*) at 4 °C and discard the supernatant.
12. Add 70 μL sterile 10% glycerol and resuspend with a pipette gently (*see* **Note 6**).

3.1.3 Transformation to Generate Transposon Mutant Library

1. Make transposome complex for transformation.
2. Mix transposon fragment (2 μL, 300 ng) from Subheading 3.1.1, transposase (4 μL, 4 U), and 2 μL of 100% sterile glycerol and incubate at room temperature for 30 min then 4 °C overnight (*see* **Note** 7).
3. Electroporate SH competent cells with 1.5 μL of transposome complex at 2450 V and immediately add 500 μL of prewarmed SOC media to the cells for recovery.
4. Gently pipet 3 or 4 times and transfer to a 1.5 tube immediately.
5. Incubate cells at 37 °C for 1 h for recovery.
6. Spread 100 μL of transformed SH cells on the LB agar plates supplemented with kanamycin and incubate at 37 °C for 18 h.
7. Store SH mutant library at −80 °C until use.

3.2 Whole-Genome Sequencing (WGS)

The WGS protocol described here is modified from a previous protocol by Rothrock et al. [13].

3.2.1 DNA Extraction

1. Pure genomic DNA is extracted from SH transposon mutant strain using a DNeasy Blood and Tissue kit according to the manufacturer's provided protocol. Detailed protocol described below.
2. Spin down 2 mL of overnight grown culture in 1.5 mL tube at 11,000 × *g* for 3 min to harvest cells and discard supernatant.
3. Add 180 μL of ATL to resuspend a pellet and then add 20 μL of proteinase K.
4. Vortex the mixture 10 s and incubate on a heat block at 56 °C for 5 min.
5. Immediately add 200 μL of each AL and 100% ethanol, and vortex 10 s.
6. Add entire volume of lysates into a spin column and centrifuge at 6000 × *g* for 1 min.
7. Replace collection tube and add 500 μL of AW1 and centrifuge at 6000 × *g* for 1 min.
8. Replace collection tube and add 500 μL of AW2 and centrifuge at 6000 × *g* for 3 min.
9. Position spin column onto a new 1.5 mL tube. Add 100 μL of DNase–RNase-free water and centrifuge at 11,000 × *g* for 1 min.
10. Keep extracted genomic DNA at −20 °C until use.

3.2.2 Qubit and Quantification

1. Genomic DNA concentration must be measured using a Qubit 4.0 Fluorometer and diluted to 0.2 ng/μL.
2. Prepare the Qubit working solution by diluting the Qubit™ reagent 1:200 in Qubit™ buffer according to the manufacturer's recommendation.
3. Standard samples 1 and 2 must be read prior to measuring samples.
4. Add 190 μL working solution to an assay tube containing 10 μL of each standard 1 and 2.
5. Vortex gently and read standards.
6. Add 198 μL working solution to an assay tube containing 2 μL of sample DNA.
7. Vortex gently and incubate for 2 min at room temperature and read samples.

3.2.3 Tagmentation and Amplification

1. Thaw ATM, TB, NPM, and index primers (1 and 2) at room temperature.
2. Add 10 μL of TB to a PCR tube and add 5 μL of sample DNA.
3. Add 5 μL of ATM, mix by pipetting and quick centrifuge the PCR tube.
4. Use a thermocycler to add adapter sequences to DNA fragment of 250 bp by transposase activity in ATM mix. Program the thermocycler to hold at 55 °C for 5 min followed by 10 °C.
5. As soon as the thermocycler reaches 10 °C, immediately add 5 μL NT to stop tagmentation reaction.
6. Mix sample by pipetting and quick centrifuge the PCR tube.
7. Incubate sample at room temperature for 5 min and add 15 μL NPM to the PCR tube.
8. Select appropriate index scheme and add 5 μL of index primer 1 and mix by pipetting.
9. Add 5 μL of index primer 2, mix by pipetting and quick centrifuge the PCR tube.
10. Run PCR as follows: 72 °C for 3 min, 95 °C for 30 s, 12 cycles of 95 °C for 10 s, 55 °C for 30 s, 72 °C for 30 s, and 72 °C for 5 min. Hold at 10 °C at the end of cycle.
11. Quick spin a tube using the microcentrifuge and hold at 4 °C up to 24 h.

3.2.4 PCR Cleanup

1. Bring AmpPure beads to room temperature by placing on bench for 30 min then vortex the beads for 1 min.
2. Add 25 μL of AmpPure beads to a 1.5 mL tube.
3. Quick spin PCR tube on the microcentrifuge and transfer 45 μL of PCR product to the 1.5 mL tube containing AmpPure beads.
4. Mix by pipetting and incubate the 1.5 mL tube at room temperature for 5 min.
5. Place the 1.5 mL tube on a magnetic stand and wait for beads to adhere to sides of well.
6. Discard supernatant when becomes clear and add 80% ethanol while on magnetic stand to wash and immediately remove ethanol from beads.
7. Remove ethanol completely by pipetting and allow to air-dry on magnet stand for 10 to 5 min.
8. Remove the 1.5 mL tube from the magnetic stand and add 52.5 μL RSB to the tube containing dry beads.
9. Mix by pipetting and incubate the 1.5 mL tube at room temperature for 2 min.

10. Place the 1.5 mL tube back on the magnet stand until supernatant becomes clear.
11. While on the magnetic stand, transfer 45 μL of supernatant to a new sterile 1.5 mL tube for storage up to 1 week at −20 °C.

3.2.5 Library Normalization

1. Prepare LNA1, LNB1, and LNW1 reagents at room temperature.
2. Prepare LNA–LNB mixture by adding 8 μL of LNB to 44 μL of LNA.
3. Prepare a 1.5 mL tube and add 45 μL of LNA–LNB mixture.
4. Add 20 μL PCR library from Subheading 3.2.4 and shake for 30 min at 300 × *g*.
5. After shaking, place the 1.5 mL tube on the magnetic stand until the supernatant becomes clear then discard supernatant.
6. Remove the 1.5 mL tube from the magnetic stand, add 45 μL of LNW and shake for 5 min at 300 × *g*.
7. Repeat **steps 5** and **6**.
8. Add 30 μL of fresh 0.1 N NaOH and shake for 5 min at 300 × *g*.
9. After shaking, transfer 30 μL of supernatant to the 1.5 mL tube containing 30 μL of LNS.
10. Centrifuge the 1.5 mL tube at 800 × *g* for 1 min.

3.2.6 Library Pooling and MiSeq Operation

1. Add 24 μL of library in HT1 to a 1.5 mL tube containing 576 μL of HT1 to make a diluted amplicon library (DAL) (*see* **Note 8**).
2. Heat DAL for 2 min at 96 °C.
3. Gently invert tube, then cool the mixture in a water–ice bath for 5 min.
4. Spike the library mixture with a PhiX V3 with a concentration (2%).
5. Load 600 μL of diluted library to Illumina MiSeq cartridge (*see* **Note 9**).
6. Upload sample sheet to the MiSeq and start the run by following prompted steps.

4 Notes

1. Components in the kit. ATL (Tissue Lysis Buffer), proteinase K, AL (Lysis Buffer), spin column, collection tube, AW1 (Wash Buffer), AW2 (Wash Buffer).

2. Components in the kit. ATM (Amplicon Tagment Mix), TB (Tagment DNA Buffer), NPM (Nextera PCR Master Mix), index primers (1 and 2), NT (Neutralize Tagment Buffer), AMPure XP beads, RSB (Resuspension Buffer), LNA1 (Library Normalization Additives), LNB1 (Library Normalization Beads 1), LNW1 (Library Normalization Wash 1), LNS (Library Normalization Storage Buffer 1).
3. This protocol is optimized for kanamycin sensitive Gram-negative bacteria.
4. Electrophoresis of the digested fragment to reduce potential mutation caused by the EtBr.
5. Always keep them on ice from this step onward, unless specified otherwise.
6. Use the SH competent cell immediately for better recovery or store at −80 °C for few days.
7. The concentration of transposon fragment should be at least 150 ng/μL.
8. If sample number is between 2 and 24, each sample should be pooled to make one. Five microliters of each sample library is added to one 1.5 mL tube and vortex pooled library thoroughly. Dilute PAL (pooled amplicon libraries) in HT1 by adding 24 μL PAL to a tube containing 576 μL of HT1 to make DAL (diluted amplicon libraries).
9. Take an Illumina reagent cartridge from −20 °C and thaw in DI water bath for 1 h.

References

1. Pray LA (2008) Transposons: the jumping genes. Nature Education 1(1):204
2. Wicker T, Sabot F, Hua-Van A, Bennetzen JL, Capy P, Chalhoub B, Flavell A, Leroy P, Morgante M, Panaud O (2007) A unified classification system for eukaryotic transposable elements. Nat Rev Genet 8(12):973
3. Finnegan DJ (1989) Eukaryotic transposable elements and genome evolution. Trends Genet 5:103–107
4. Thermo Fisher Scientific (2018) Introduction to transfection. https://www.thermofisher.com/us/en/home/references/gibco-cell-culture-basics/transfection-basics/introduction-to-transfection.html. Accessed 6 Sep 2018
5. Nguyen DH (2018) Four ways to insert foreign DNA into cells. Hearst Seattle Media, LLC, Seattle, WA. https://education.seattlepi.com/four-ways-insert-foreign-dna-cells-4064.html. Accessed 6 Sep 2018
6. Rivera AL, Gómez-Lim M, Fernández F, Loske A (2014) Genetic transformation of cells using physical methods. J Genet Syndr Gene Ther 5(4):237–242
7. Freed NE (2017) Creation of a dense transposon insertion library using bacterial conjugation in enterobacterial strains such as *Escherichia coli* or *Shigella flexneri*. J Vis Exp (127)
8. Voelkerding KV, Dames SA, Durtschi JD (2009) Next-generation sequencing: from basic research to diagnostics. Clin Chem 55(4):641–658
9. Van Opijnen T, Camilli A (2013) Transposon insertion sequencing: a new tool for systems-level analysis of microorganisms. Nat Rev Microbiol 11(7):435
10. Barquist L, Boinett CJ, Cain AK (2013) Approaches to querying bacterial genomes with transposon-insertion sequencing. RNA Biol 10(7):1161–1169

11. Kwon YM, Ricke SC, Mandal RK (2016) Transposon sequencing: methods and expanding applications. Appl Microbiol Biotechnol 100(1):31–43
12. Khatiwara A, Jiang T, Sung S-S, Dawoud T, Kim JN, Bhattacharya D, Kim H-B, Ricke SC, Kwon YM (2012) Genome scanning for conditionally essential genes in *Salmonella*. Appl Environ Microbiol 78:3098–3107.13
13. Rothrock MJ, Fan P, Jeong KC, Kim SA, Ricke SC, Park SH (2018) Complete genome sequence of *Listeria monocytogenes* strain MR310, isolated from a pastured-flock poultry farm system. Genome Announc 6(10): e00171–e00118

Chapter 3

Construction of a Sequence-Defined Transposon Mutant Library in *Staphylococcus aureus*

Jennifer L. Endres, Vijaya Kumar Yajjala, Paul D. Fey, and Kenneth W. Bayles

Abstract

Transposon mutagenesis is one of the most widely used strategies to generate a large number of random mutations within a bacterial genome and then to precisely identify the mutated sites. The generation of sequence-defined transposon mutant libraries that are composed of a collection of different mutants, each containing a single transposon insertion mutation within nearly all of the nonessential genes within the genome, is a rapid and reliable way to enhance the study of gene function. In this chapter, we describe the process to generate a sequence-defined transposon mutant library in *Staphylococcus aureus* utilizing the mariner-based *bursa aurealis* transposon.

Key words *Staphylococcus aureus*, Transposon, Random mutagenesis, *bursa aurealis*

1 Introduction

Staphylococcus aureus is a commensal organism in the nares of roughly 30% of the human population [1]. The genome encodes a vast array of virulence factors that allow it to evade the host immune response and cause severe disease in almost all niches within the human body. In addition to this versatile pathogenic capacity, *S. aureus* has become resistant to almost all current anti-bacterial treatments and continues to be a major cause of morbidity and mortality worldwide [2]. Quick access to transposon mutant libraries has allowed investigators to probe and understand the complex strategies used by *S. aureus* to overcome the host immune system, as well as to understand the molecular mechanisms of antibiotic resistance.

Transposons integrate randomly into genomes typically causing inactivity of the genes in which they insert. If the process is allowed to occur enough times, saturated "pools" of mutants will be generated where each individual contains a single random

Steven C. Ricke et al. (eds.), *Microbial Transposon Mutagenesis: Protocols and Applications*, Methods in Molecular Biology, vol. 2016, https://doi.org/10.1007/978-1-4939-9570-7_3,

transposon insertion. Ideally, this collection of mutants will be diverse enough that it represents the majority of the nonessential genes within the genome. Isolation and then sequencing of the "junction fragments" containing the interface between the transposon and the genome allows for the precise identification of each of the transposon insertion sites.

Instead of pooling these mutants and subjecting the entire population to screening or selection approaches, it is valuable to array the individual mutants into a format where each insertion site has been determined by DNA sequencing. These "sequence-defined" transposon mutant libraries allow researchers quick access to mutants and are ideally suited for phenotypic screens or to test hypotheses generated as a result of a variety of "omics" studies. These libraries have been generated in many different bacterial species, including *Vibrio cholera* and *Pseudomonas aeruginosa* [3–5].

Recently, the *bursa aurealis* transposon [6–8] was used to generate the "Nebraska Transposon Mutant Library," which is a collection of 1,952 sequence-defined mutants that represent the majority of the nonessential genes in the *S. aureus* genome [5]. In this chapter, we describe the steps required to generate sequence-defined transposon mutants utilizing the *bursa aurealis* system in *S. aureus* (Fig. 1).

The following process is adapted from the previously reported procedures [9, 10].

2 Materials

Prepare all reagents using ultrapure water and analytical grade reagents. Prepare and store all reagents at room temperature (unless indicated otherwise).

2.1 Components Required for Heat Shock and Patching

1. Tryptic soy agar (TSA).
2. Incubator and Heat block set at 45 °C.
3. Toothpicks.
4. Antibiotics: Chloramphenicol (10 mg/mL), erythromycin (25 mg/mL), and tetracycline (5 mg/mL).
5. 3 mm Glass beads.
6. 1.5 mL epitubes.

2.2 Components for Genomic DNA Isolation

1. Tryptic soy broth (TSB).
2. Erythromycin (5 mg/mL).
3. Shaking incubator set at 37 °C.
4. 50% glycerol.

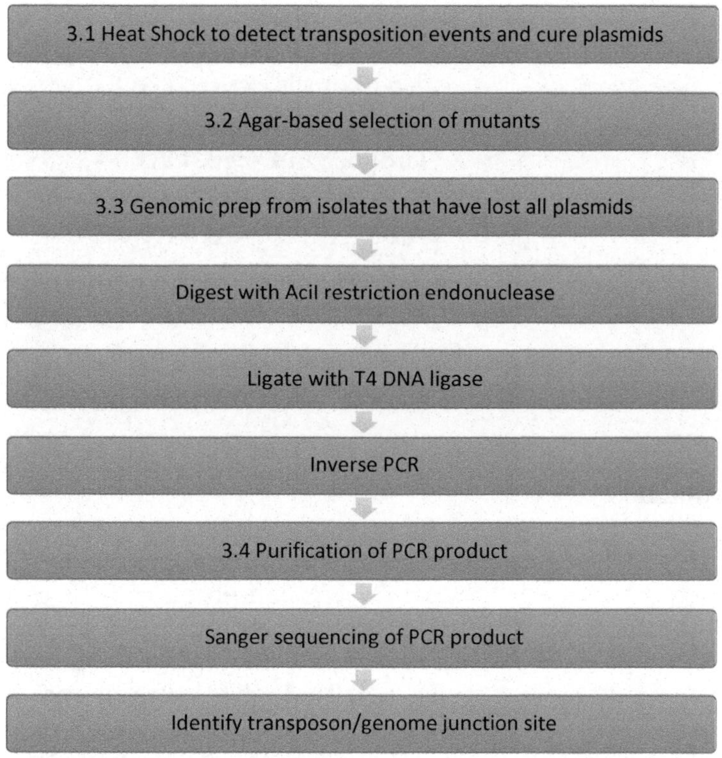

Fig. 1 Flowchart of the steps required for the generation of random transposition events in *Staphylococcus aureus*

5. Colony Picker (cat# VP 373, V & P Scientific, Inc.) or toothpicks.
6. 1 mL 96-well polypropylene plate.
7. 2 mL 96-well polypropylene plate.
8. Wizard Genomic DNA purification Kit (Promega, Madison, WI).
9. 50 mM EDTA.
10. Lysostaphin (Ambi Products).
11. Oven.
12. Tris–EDTA Buffer. (For 100 mL; 1 mL of 1 M Tris (pH 8), 0.5 mL of 0.5 M EDTA (pH 8) + 98.5 mL water.)
13. 70% ethyl alcohol.
14. Isopropyl alcohol.

2.3 Molecular Genetic Components to Confirm Transposon Insertion Site

1. 96-well PCR plates.
2. Semiskirted 96-well PCR plates.
3. AciI restriction enzyme.

4. Ligation master mix. (For each sample; 2.5 μL T4 DNA ligase buffer, 0.5 μL dilution buffer, 0.5 μL T4 DNA ligase, and 1.5 μL nuclease-free H_2O.)
5. Taq DNA polymerase and buffers (Monserate Biotechnology Group, San Diego, CA).
6. Forward primer (Buster) 5′-GCTTTTTCTAAATGTTTTTT AAGTAAATCAAGTACC-3′.
7. Reverse primer (Martn ermR) 5′-AAACTGATTTTTAGTAAA CAGTTGACGATATTC-3′.
8. Thermocycler.
9. ExoSAP-IT (Affymetrix, Santa Clara, CA).

3 Methods

First, plasmids containing a transposase mutant derivative of the *bursa aurealis* transposon and the gene encoding the transposase (designated pBursa and pFA545, respectively; *see* ref. [11]) are introduced into the *S. aureus* strain in which the library will be generated. To do this, bacteriophage Φ11 transducing lysates generated on *S. aureus* RN4220 carrying pFA545 is used to transduce this plasmid into the desired *S. aureus* strain and plated on selective media (tetracycline at 5 μg/mL) where they are allowed to grow overnight at 30 °C. Once the presence of pFA545 is confirmed, this strain is used as a recipient in a second transduction to introduce the pBursa plasmid and plating on selective media for both plasmids (tetracycline and chloramphenicol), followed by another overnight incubation at 30 °C. Once the strain that harbors both plasmids is confirmed, continue with the heat-shock protocol below (*see* **Note 1**).

3.1 Heat Shock to Detect Transposition Event and Cure Plasmids pBursa and pFA545

1. Aliquot 1 mL of sterile water in 1.5 mL microcentrifuge tubes.
2. Place tubes in a heat block at 45 °C for 1 h.
3. With a sterile toothpick, pick a single transductant per tube and disperse to make the water turbid.
4. Plate 100–200 μL of the sample onto prewarmed TSA plates containing 25 μg/mL erythromycin.
5. Incubate plates at 45 °C for 2 days.

3.2 Agar-Based Selection of Mutants

Agar-based selection is used to confirm whether the plasmids have been successfully cured from the possible transposon mutants (*see* **Note 2**).

1. Patch the colonies obtained from the heat shock procedure onto three TSA plates each containing a different antibiotic

(25 μg/mL erythromycin, 10 μg/mL chloramphenicol and 5 μg/mL tetracycline).

2. Consider the patches that grow only on TSA plates containing erythromycin, which indicate that they have been cured of both plasmids.

3.3 Identification of Random Transposon Insertion Sites by Inverse PCR

3.3.1 Isolation of Genomic DNA by Modified Promega Wizard Genomic Prep Protocol

Day 1:

1. Aliquot 400 μL of TSB containing 5 μg/mL erythromycin to each well of a 1 mL 96-well plate.
2. Using a Colony Picker or toothpicks, inoculate one mutant into each well containing TSB with erythromycin (5 μg/mL) and shake at 250 rpm overnight at 37 °C.

Day 2:

1. Centrifuge the 96-well plate to pellet the cells (3000 × *g* for 10 min).
2. Discard the supernatant and resuspend pellets in 110 μL of 50 mM EDTA.
3. Add 10 μL of a 10 mg/mL solution of lysostaphin and mix vigorously until an evenly distributed cell suspension is obtained.
4. Incubate at 37 °C for 90 min (mixture should become viscous and translucent).
5. Add 600 μL of Promega Nuclei Lysis Buffer and incubate the plate at 80 °C for 10 min.
6. Cool to room temperature and then add 200 μL of Promega Protein precipitation solution. Vortex vigorously for 2 min and then place on ice for 10 min.
7. Centrifuge at 3000 × *g* for 10 min.
8. Transfer the supernatant (without disturbing the pellet) into a 2 mL 96-well plate containing 600 μL isopropanol. Mix well by inverting.
9. Centrifuge at 3000 × *g* for 10 min to collect the precipitated DNA.
10. Discard the supernatant, being careful not to disrupt the DNA pellet.
11. Add 600 μL of room temperature 70% ethanol and invert 5 to 10 times to wash the pellet. Centrifuge at 3000 × *g* for 10 min.
12. Discard the ethanol wash and dry the pellet by leaving the plate open for 10 to 15 min; make sure all the ethanol is evaporated.
13. Rehydrate the DNA pellet in 100 μL TE buffer and incubate the plate at 65 °C for 1 h.
14. Store the genomic DNA at −20 °C until further use.

3.3.2 Digestion of Genomic DNA and Ligation

Day 3:

1. Digests are performed in 20 μL total volume in a 96-well PCR plate (17 μL genomic DNA, 2 μL New England Biolabs Buffer 2, 1 μL New England Biolabs AciI restriction endonuclease).
2. Incubate at 37 °C for 2 h. Heat-inactivate the enzyme at 65 °C for 20 min.
3. Aliquot 5 μL of the ligation master mix into each reaction and mix (total volume of 25 μL).
4. Incubate overnight at 4 °C.

3.3.3 Inverse PCR Protocol and Purification of Amplified DNA

Day 4:

Inverse polymerase chain reaction (PCR) is performed to identify the transposon insertion site within the genome using primers that anneal to two different loci on the transposon. After AciI digestion and ligation with T4 DNA ligase, the genome is a collection of circular DNA molecules of various sizes. Some of these molecules will contain *bursa aurealis*, thereby supplying a known primer binding site to bind and amplify the entire molecule by inverse PCR and providing a template for sequencing the transposon–genome junction site.

1. Make the following master mix for each sample (5 μL 10× PCR buffer, 1 μL Taq polymerase, 1 μL 10 μM Buster primer, 1 μL 10 μM martn erm R primer, 1 μL dNTP, 1 μL 50 mM $MgCl_2$, and 35 μL nuclease-free water).
2. Aliquot 45 μL of the master mix into each well of a 96-well PCR plate.
3. Add 5 μL of the AciI digested and T4 DNA ligated genomic DNA.
4. Amplify the products in a thermocycler with the following cycles (40 cycles of 94 °C for 30 s, 63 °C for 30 s, and 72 °C for 3 min).
5. Analyze 10 μL of the PCR products in a 1% agarose gel to check for the generation of random sized PCR products as illustrated in Fig. 2, indicative of random transposon insertion.

3.4 Purification of Inverse PCR Fragments and Sequencing of bursa aurealis-Genome Insertion Sites

Once the banding pattern of your PCR fragments has confirmed random insertion of the transposon, enzymatic cleanup of your fragments before sequencing is performed as outlined below.

1. Transfer 6 μL if the PCR products from each well of the 96-well plate to the corresponding well of a 96-well sequencing plate (semiskirted).
2. Add 2 μL of EXOSAPIT, mix well, and incubate at 37 °C for 15 min.

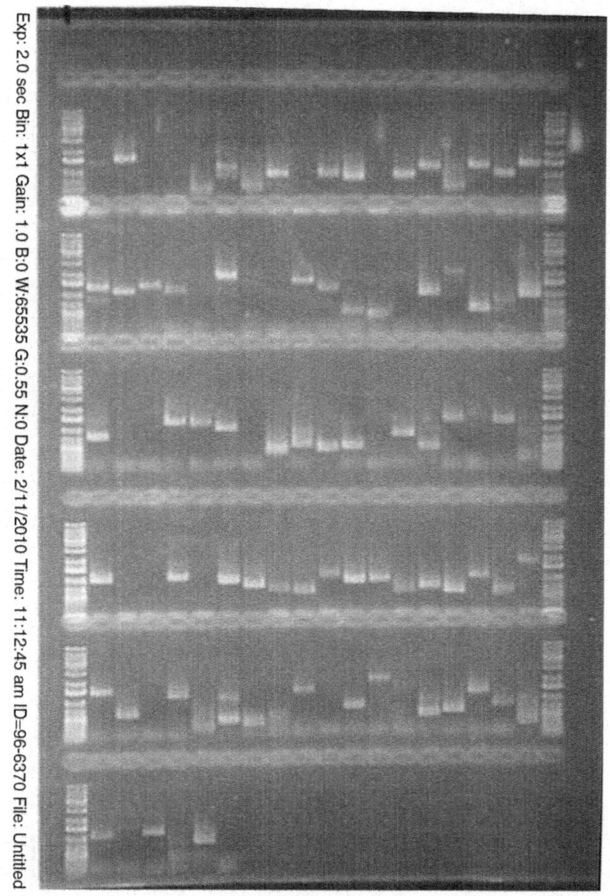

Fig. 2 Random banding pattern of 96 PCR reactions in a 1% agarose gel

3. Heat-inactivate the enzyme at 80 °C for 15 min.
4. Allow the samples to cool to room temperature and then use the Buster primer (5′-GCTTTTTCTAAATGTTTTTTAAGT AAATCAAGTACC-3′) for sequencing of the transposon–genome junction site.
5. Once high-quality sequencing is obtained, identify the transposon substring of CCTGTTA using a sequence viewer (*see* **Note 3**).
6. Copy the next 100–200 bp of sequencing results making sure not to pass an AciI site (CCGC), which suggests that you have reached the other end of the transposon.
7. Using the NCBI BLAST site, blast the sequence and then scroll down to find the reference strain most closely related to your strain.

8. Collect the following information into an excel document.
   (a) Transposon insertion site coordinates.
   (b) The gene or intergenic region of the chromosome with which the transposon has inserted.
   (c) The orientation of the transposon and the gene, as well as the start and end coordinates of the gene (*see* **Note 4**).

4 Notes

1. To avoid the generation of clonal populations, the pBursa plasmid is freshly transduced into the strain carrying pFA545 every time a heat shock is performed to generate new mutants. Also, more transposition events are detected from each heat shock when the transductant colonies (carrying both pFA545 and pBursa) have been able to incubate at 4 °C for 5–7 days.
2. Typically, 90% of colonies are both tetracycline and chloramphenicol sensitive suggesting that successful curing of both of the plasmids has taken place.
3. When the Buster primer is used for sequencing, the *bursa aurealis* substring (CCTGTTA) is typically identified 65–75 nucleotides into the sequencing reads.
4. If the transposon and gene are in the same orientation, the gene encoding green fluorescent protein (*gfp*) present on *bursa aurealis* can possibly be used as a transcriptional fusion for that gene.

References

1. Gorwitz RJ, Kruszon-Moran D, McAllister SK et al (2008) Changes in the prevalence of nasal colonization with *Staphylococcus aureus in the United States, 2001-2004.* J Infect Dis 197 (9):1226–1234
2. Archer GL (1998) *Staphylococcus aureus:* a well-armed pathogen. Clin Infect Dis 26 (5):1179–1181
3. Cameron DE, Urbach JM, Mekalanos JJ (2008) A defined transposon mutant library and its use in identifying motility genes in *Vibrio cholerae.* Proc Natl Acad Sci U S A 105 (25):8736–8741
4. Jacobs MA, Alwood A, Thaipisuttikul I et al (2003) Comprehensive transposon mutant library of *Pseudomonas aeruginosa.* Proc Natl Acad Sci U S A 100(24):14339–14344
5. Fey PD, Endres JL, Yajjala VK et al (2013) A genetic resource for rapid and comprehensive phenotype screening of nonessential *Staphylococcus aureus* genes. MBio 4(1): e00537–e00512. https://doi.org/10.1126/mBio.00537-12
6. Tam C, Glass EM, Anderson DM et al (2006) Transposon mutagenesis of *Bacillus anthracis* strain sterne using *Bursa aurealis.* Plasmid 56 (1):74–77
7. Bae T, Banger AK, Wallace A et al (2004) *Staphylococcus aureus* virulence genes identified by *bursa aurealis mutagenesis* and nematode killing. Proc Natl Acad Sci U S A 101 (33):12312–12317
8. Bae T, Glass EM, Schneewind O et al (2008) Generating a collection of insertion mutations in the *Staphylococcus aureus* genome using *bursa aurealis.* Methods Mol Biol 416:103–116
9. YajjalaVK, Widhelm TJ, Endres JL et al (2016) Generation of a transposon mutant library in *Staphylococcus aureus* and *Staphylococcus*

epidermidis using bursa aurealis. Methods Mol Biol 1373:103–110

10. Widhelm TJ, Yajjala VK, Endres JL et al (2014) Methods to generate a sequence-defined transposon mutant library in *Staphylococcus epidermidis* strain 1457. Methods Mol Biol 1106:135–142
11. Krausz KL, Bose JL (2016) Bacteriophage transduction in *Staphylococcus aureus*: broth-based method. Methods Mol Biol 1373:63–68

Chapter 4

Transposon Mutagenesis in *Streptococcus* Species

Martin Nilsson, Michael Givskov, and Tim Tolker-Nielsen

Abstract

Mutant libraries, generated by transposons and screened for various phenotypes, have led to many important discoveries regarding gene functions in various organisms. In this chapter we describe the use of plasmid pMN100, a transposon vector constructed to perform in vivo transposition primarily in oral streptococci. Compared to in vitro transposition systems the conditional replicative features of the plasmid, and the inducible expression of the mariner *Himar1* transposase, makes pMN100 particularly useful for bacterial strains showing a low transformation frequency. We outline how to transform plasmid pMN100 into *Streptococcus mutans*, carry out transposon mutagenesis, and determine the chromosomal location of inserted transposons. It is our prospect that the protocols can be used as guidelines for transposon mutagenesis in *S. mutans* as well as other species of streptococci.

Key words In vivo transposon mutagenesis, Mariner, pMN100, Streptococci

1 Introduction

Transposon mutagenesis has been essential for gene discovery in bacteria. A successful outcome of a transposon mutant library screen is dependent on several factors, where two crucial parameters are the diversity of the library and the number of mutants screened. Many different transposon gene delivery approaches have been used to create mutant libraries in bacteria and there is a wealth of articles describing various transposon systems [1]. This information in combination with studies exploring the basic mechanisms of various transposons forms a solid base in order to consider and select an attractive transposition strategy. We requested a transposon system that was optimal for oral streptococci. Tn916, Tn917 and ISS1 are examples of transposons that have been utilized for this group of bacteria [2–4]. Although interesting studies have been conducted with Tn916 and Tn917, these transposons have been shown to possess some features that limit the efficiency of constructing an unbiased transposon library. Tn916 has preferred insertion sites consisting of adenine rich sequences separated by six

Steven C. Ricke et al. (eds.), *Microbial Transposon Mutagenesis: Protocols and Applications*, Methods in Molecular Biology, vol. 2016, https://doi.org/10.1007/978-1-4939-9570-7_4,

bases from thymine-rich sequences [1]. Tn917 has been reported, in some bacteria, to be inserted nonrandomly in the chromosome, and instead occur in a much higher frequency in certain regions [5]. The ISS1 delivery vectors, such as pGh9:ISS1, mediate trans position through a replicative mechanism, where the whole plasmid or multiple plasmid sequences are integrated into the chromosome. In addition, some bacterial strains harbor endogenous copies of ISS1, which can potentially form targets for recombination events [6].

After considering these options thoroughly, we turned our attention to the *mariner* transposon system. Transposons of the *mariner* family are widespread among eukaryotes [7]. As a tool for *mariner*-based mutagenesis in bacteria, a variant of the *Himar1* transposon, originally isolated from the horn fly *Haematobia irritans*, has become a common choice [8]. There is extensive data showing that transposition based on the *mariner* transposons is essentially unbiased [9]. The recognition sequence of *mariner* transposons is only two base pairs (TA dinucleotides). In addition *mariners* do not require any host factors. Due to the latter feature the transposition can be performed in vitro [10], but to generate a sufficient number of transposon mutants, the in vitro system requires a strain that is readily transformable. A challenge among oral streptococci is that several strains show a low frequency of transformation. Therefore, we employ an in vivo transposition strategy with a temperature-sensitive plasmid comprising an inducible *Himar1* transposase. This means that theoretically only one transformant harboring the plasmid has to be obtained, and the construct can subsequently be propagated before transposition is induced and transposon mutants are generated.

In order to construct a transposon vector with the key features described above we fused the transposon-expressing cassette from pBTn with the backbone fragment of pTV1-OK [3, 11]. The transposon vector was termed pMN100 and is schematically shown in Fig. 1 [12]. The kanamycin resistance gene *aphA3* and the temperature-sensitive origin of replication *repAts* in pMN100 originate from pTV1-OK, a transposon vector with Tn917 used in streptococci. Both the kanamycin resistance and the *repAts* segment are functional in *Escherichia coli* as well as Gram-positive hosts including streptococci. The transposon-expressing cassette derived from pBTn contains an erythromycin resistance gene flanked by *Himar1* inverted repeats, and the *Himar1* transposase gene under control of a xylose inducible promoter. A *S. mutans* transposon library obtained with pMN100 has successfully been used to study genes involved in biofilm-associated antimicrobial tolerance [13]. The following protocols describe the use of pMN100 for construction of a transposon mutant library in *S. mutans*, that is, transformation of plasmid pMN100 into the target strain, transposon mutagenesis, as well as the subsequent identification of the chromosomal location of transposon inserts (Fig. 2).

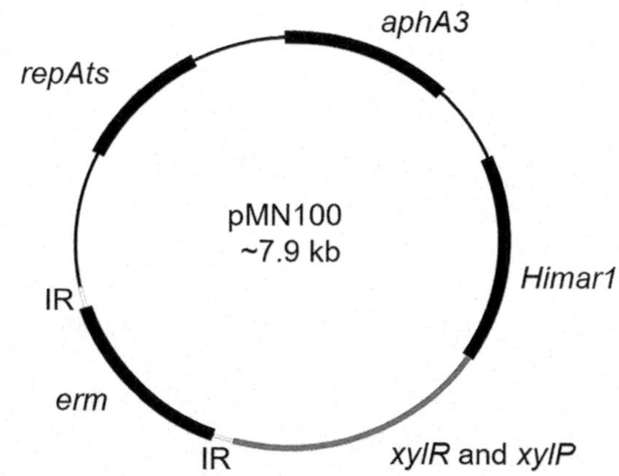

Fig. 1 Schematic of the *Himar1*-based transposon vector pMN100. The plasmid contains a temperature-sensitive broad host range origin of replication (*repAts*-pWV01), a kanamycin resistance gene (*aphA3*), an erythromycin resistance gene (*erm*) flanked with mariner inverted repeats (IR), the mariner *Himar1* transposase gene under the control of the xylose inducible promoter (*xylP*), as well as the xylose regulator gene (*xylR*)

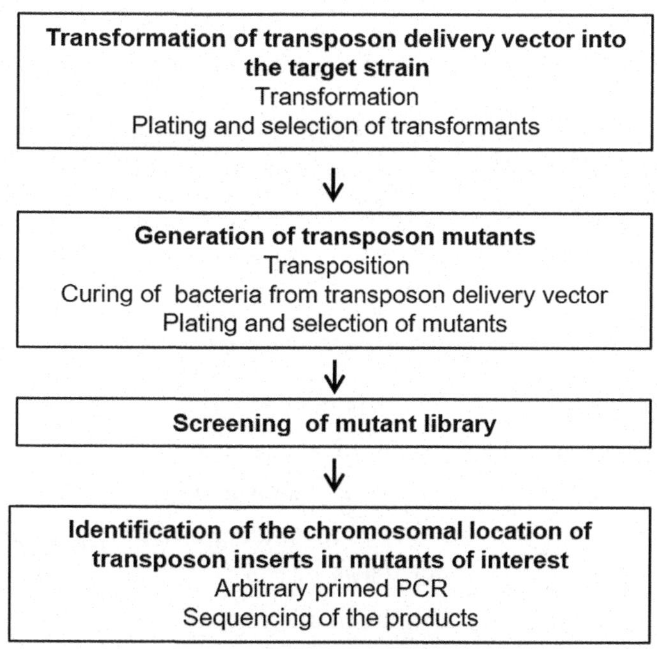

Fig. 2 Gene discovery in bacteria often involves transposon mutagenesis, mutant library screening, and identification of the transposon insertion site in mutants of interest. The present protocol describes transformation of a *Himar 1* transposon delivery vector into *Streptococcus mutans*, *Himar 1*-based in vivo transposon mutagenesis, and a procedure to locate the transposon insertion sites in mutants of interest

2 Materials

2.1 Growth Media

Media are prepared with deionized water, autoclaved, and stored at room temperature.

1. Tryptone soya agar (TSA) from Oxoid consisting of pancreatic digest of casein 15 g/l, enzymatic digest of soya bean 5 g/l, sodium chloride 5 g/l, and agar 15 g/l. Dissolve 40 g in 1 l. After autoclaving, antibiotic, if required, is added to the cooled (55 °C) medium.
2. Bacto™ Todd Hewitt broth (THB) from BD consisting of heart infusion from 500 g 3.1 g/l, neopeptone 20.0 g/l, dextrose 2 g/l, disodium chloride 2 g/l, and sodium carbonate 2.5 g/l. Dissolve 30 g in 1 l. pH is adjusted to 7.6
3. Tryptone soya broth (TSB) from Oxoid consisting of pancreatic digest of casein 17.0 g/l, enzymatic digest of soya bean 3.0 g/l, sodium chloride 5.0 g/l, dipotassium hydrogen phosphate 2.5 g/l, and glucose 2.5 g/l. Dissolve 30 g in 1 l.
4. Bacto™ Tryptone soya broth without dextrose from BD consisting of pancreatic digest of casein 17.0 g/l, enzymatic digest of soya bean 3.0 g/l, sodium chloride 5.0 g/l, and dipotassium hydrogen phosphate 2.5 g/l. Dissolve 27.5 g in 1 l.

2.2 General Equipment and Basic Chemicals

1. Pipettes and sterile 20, 100, and 1000 μl pipette tips.
2. Eppendorf tubes (1.5 ml).
3. Petri dishes.
4. 50 mg/ml kanamycin dissolved in MQ-H_2O (sterile filtered).
5. 10 mg/ml erythromycin dissolved in ethanol.
6. Anaerobic box.

2.3 Transformation

1. Plasmid pMN100.
2. An *S. mutans* strain.
3. Heat-inactivated horse serum.
4. Competence stimulating peptide (CSP) for *S. mutans.*
5. Glycerol.
6. Incubation tubes (15 ml).
7. Cryo tubes.
8. Autoclave.
9. NanoDrop spectrophotometer.
10. Spectrophotometer.
11. Tabletop centrifuge.

2.4 Transposon Mutagenesis

1. *S. mutans*/pMN100.
2. Tabletop centrifuge.
3. 96-well microtiter plates.
4. Incubation tubes (15 ml).
5. Cryo tubes.
6. Glycerol.
7. Sterile toothpicks.
8. 96-pin replicator.
9. Plastic covers for microtiter plates.
10. −80 °C freezer.
11. Xylose.
12. 100 ml Erlenmeyer flask.

2.5 Identification of Transposon Insertion Site

1. Taq DNA polymerase from *Thermus aquaticus*.
2. dNTP mix (2.5 mM/nucleotide).
3. 10× PCR buffer.
4. Dimethylsulfoxide (DMSO).
5. Ethidium bromide solution 0.07%.
6. 6× Loading Dye.
7. 1× Tris–acetate–EDTA solution (TAE). Diluted in water from a 50× solution.
8. DNA ladder.
9. PCR purification kit.
10. DNA purification kit.
11. Gel electrophoresis equipment.
12. UV table.
13. PCR thermocycler.
14. PCR tubes.
15. Agarose.
16. Microcentrifuge for PCR tubes.
17. Primers (*see* Table 1).

3 Methods

***3.1 Transformation of* S. mutans**

1. Plate *S. mutans* on TSA and incubate at 37 °C anaerobically until colonies appear.
2. Inoculate 10 ml of THB with a single colony (*see* **Note 1**) and incubate overnight at 37 °C.

Table 1
Oligonucleotides used in the protocol

Primer name	Sequence
erm 5.1	GCTTCTAAGTCTTATTTCCATAAC
erm 5.2	AGATAATGCACTATCAACACACTC
erm 5.3	TCTACATTACGCATTTGGAATAC
erm 3.1	TAGGTATACTACTGACAGCTTC
erm 3.2	ATTCTATGAGTCGCTTTTGTA
erm 3.3	TACTTATGAGCAAGTATTGTCTA
arb1	GGCCACGCGTCGACTAGTCANNNNNNNNNNGATAT
arb2	GGCCACGCGTCGACTAGTCANNNNNNNNNNGATCA
arb3	GGCCACGCGTCGACTAGTCA

3. Transfer 200 μl overnight culture to 10 ml prewarmed THB containing 5% heat-inactivated horse serum (*see* **Note 2**). The optical density (OD) of the culture is measured at 600 nM on a spectrophotometer (*see* **Note 3**). At approximately OD 0.2 the culture is left at room temperature for 10 min. In an Eppendorf tube, mix 1 ml culture with 2 μl (500 ng) CSP and 1 μg of pMN100 and incubate at 30 °C anaerobically for 60 min (*see* **Note 4**) [14].
4. Harvest cells by centrifugation at 6000 × *g* and resuspend in 200 μl THB.
5. Spread 20 μl cell suspension (add 80 μl THB to facilitate spreading), and 180 μl cell suspension on TSA plates containing 2 mg/ml kanamycin and 5 μg/ml erythromycin (*see* **Note 5**). Incubate the plates at 30 °C anaerobically for 3 days.
6. Grow separate colonies from the transformation at 30 °C in 10 ml TSB containing 300 μg/ml and 5 μg/ml erythromycin. Make culture stocks by adding 50% glycerol in TSB to the overnight cultures giving a final concentration of 15% glycerol. Freeze the stocks in cryo tubes and store at −80 °C.

3.2 Transposon Mutagenesis

1. Plate *S. mutans*/pMN100 on TSA containing 300 μg/ml kanamycin and 5 μg/ml erythromycin, and incubate at 30 °C anaerobically until colonies appear.
2. Use one or more colonies to inoculate 10 ml of TSB containing 300 μg/ml kanamycin and 5 μg/ml erythromycin. Incubate statically overnight at 30 °C.
3. Transfer 1 ml of the overnight culture to 9 ml dextrose-free TSB containing 0.5% xylose, 300 μg/ml kanamycin, and 5 μg/ml erythromycin (*see* **Note 6**). Incubate statically at 30 °C overnight.

4. Transfer 5 ml of culture to 95 ml TSB containing 5 μg/ml erythromycin. Incubate at restrictive temperature overnight (*see* **Note** 7).
5. Plate different volumes from the culture, 200 μl, 20 μl, and 2 μl (*see* **Note 8**) on TSA with 10 μg/ml erythromycin and incubate at 37 °C anaerobically until colonies appear.
6. Pick colonies from the transposition plates to a TSA plate containing 10 μg/ml erythromycin and a TSA plate containing 300 μg/ml kanamycin (*see* **Note 9**). Incubate at 37 °C anaerobically (*see* **Notes 10–12**).
7. To store mutants, pick colonies with an erythromycin-resistant and kanamycin-sensitive phenotype to 96-well microtiter plates containing 100 μl of TSB and 5 μg/ml erythromycin and grow overnight at 37 °C.
8. Add 100 μl of TSB containing 30% glycerol to each well to give a final glycerol concentration of 15%.
9. Cover plates with plastic film and keep at −80 °C for future screenings (*see* **Note 13**).

3.3 Identification of Transposon Insertion Sites by Arbitrary Primed PCR

After screening of the mutant library the chromosomal location of the transposon in interesting mutants can be identified by the two-step PCR procedure described below.

1. The reagent compositions of the first arbitrary PCR reaction master mix for one reaction (total volume of 25 μl) is shown in Table 2 (*see* **Note 14)** Scale up master mix to an appropriate volume (number of samples plus one extra).
2. Start with PCR grade water and subsequently add the reagents. Add the Taq DNA polymerase last.
3. Aliquot master mix into PCR reaction tubes, followed by template. Close the PCR tubes with a lid.
4. Vortex tubes gently and pulse-centrifugate to collect the liquid in the bottom of the tubes.
5. Run the DNA amplification on a PCR thermocycler with the following program [11]: Initial 5 min at 95 °C followed by six cycles of 94 °C 30 s, 30 °C 30 s, 72 °C 60 s and by 30 cycles of 94 °C 30 s, 45 °C 30 s, 72 °C 60 s, ending with a final elongation at 72 °C for 5 min and 4 °C forever.
6. Prepare master mix for the second PCR according to Table 3 and the guidelines above.
7. Apply the following thermocycler parameters: 30 cycles of 94 °C for 30 s, 45 °C for 30 s, and 72 °C for 30 s followed by a final elongation at 72 °C for 5 min and 4 °C forever.
8. Run approximately one-tenth of the amplified fragments on a 1% agarose gel (*see* **Note 15**).

Table 2
Composition of master mix for the first PCR

Master mix first PCR	Volume (μl)
DNA	1–5 (100–200 ng)
10× PCR buffer	2.5
dNTP mix (2.5 mM/NT)	2
arb1 (20 μM)	2
erm 5.3, 5.2, 3.3, or 3.2 (20 μM)	1
DMSO	1.25
MQ-H_2O	10.5–14.5
$MgCl_2$ (25 mM)	0.5
Taq DNA polymerase	0.25
Volume	**25**

Table 3
Composition of master mix for the second PCR

Master mix second PCR	Volume (μl)
Product from first PCR	3
10× PCR buffer	2.2
dNTP mix (2.5 mM/NT)	2
arb 3 (20 μM)	2
erm 5.2, 5.1, 3.2, or 3.1 (20 μM)	1
DMSO	1.25
MQ-H_2O	12.8
$MgCl_2$ (25 mM)	0.5
Taq DNA polymerase	0.25
Volume	**25**

9. Purify the PCR products with a PCR cleanup system following the manufacturer's instructions.
10. Sequence the products with the internal erm primer used in the second PCR.

4 Notes

1. Alternatively, a frozen aliquot (−80 °C) is used for inoculation of the start culture.
2. Thaw a tube of filter-sterilized heat-inactivated horse serum (stored at −20 °C) and add to the medium just before use. Heat-inactivated horse serum can be bought, alternatively incubate horse serum at 56 °C for 30 min.
3. The start culture should be diluted so that the requested OD600 versus blank is reached after approximately 3 h. Read OD approximately after 1.5 h growth, following OD readings each half an hour.
4. Longer incubation time will likely increase transformation frequency compared to 1 h. However, in principle, only one transformant harboring pMN100 is enough to proceed to the transposon mutagenesis procedure.
5. A higher frequency of transformants will be obtained, by plating transformation aliquots on TSA containing only one of the antibiotics. Subsequently, the transformants can be tested for the other resistance marker.
6. Xylose will induce the expression of *Himar1* transposase, which will mediate transposition of the transposon, that is, the erythromycin gene and the flanking repeats, into the chromosome of the target strain.
7. Growth at restrictive temperature will inhibit replication of pMN100 and promote the loss of plasmid, due to the temperature sensitive replicon *repAts*, and subsequently generate clones cured of pMN100. A restrictive temperature is considered to be between 40 °C and 45 °C. The temperature used is determined by the highest temperature at which the bacterial strain is able to grow well. Our *S. mutans* UA159 strain was incubated at 40 °C.
8. To get effective spreading of cells on plates, add approximately 100 μl fresh TSB medium to the 20 μl and 2 μl aliquots.
9. Use sterile toothpicks for transfer of bacteria to agar plates. Change toothpick after each transferred clone, but not between transferring to the two different agar plates. A numbered template behind each agar plate is important to use for keeping track of the order of the picked bacteria. It is usually advantageous to avoid mutants with impaired growth, that is, not to pick the smallest colonies.
10. A high degree of erythromycin- and kanamycin-resistant clones indicates that curing of the bacteria from plasmid pMN100 has not been effective. For more efficient plasmid curing, increase

the number of growth generations at restrictive temperature by diluting the culture 100× instead of 20× (i.e., 1 ml in 99 ml TSB 10 μg/ml erythromycin instead of 5 ml in 95 TSB 10 μg/ml erythromycin) after **step 3**. If plasmid curing is still not sufficient dilute the culture 100× after **step 4** and repeat the growth overnight at restrictive temperature.

11. It is important to analyze the diversity of transposon insertions, before continuing to generate a mutant library for screening. Determine the transposon insertion sites of approximately 20 mutant clones according to the protocol outlined below. If siblings occur, increase the number of separate transpositions in **step 3** [12]. How many separate transpositions that should be performed depend on the insertion frequency and the number of unique mutants one is aiming for in the library.
12. Due to a relatively low transposition frequency double transposon insertion is usually not occurring frequently.
13. For subsequent screening of the library, a 96-pin replicator can be used to spot on TSA plates (14 cm diameter petri dishes) containing 10 μg/ml erythromycin. Upon anaerobic incubation of the plates at 37 °C macrocolonies will appear, and bacteria from these can be used for specific screens.
14. Chromosomal DNA from *S. mutans* can be purified using Qiagen's DNeasy® Blood & Tissue kit according to manufacturer's instructions. Include a negative control, which contains all of the components of the PCR master mix but no template. Primers are denoted according to their location. erm 3.3, 3.2, and 3.1 bind at the 3′ end of the erythromycin gene fragment, erm 3.1 binds closest to the end followed by erm 3.2 and erm 3.3. They read out of the erythromycin fragment into the chromosomal region of the mutant. erm 5.3, 5.2, and 5.1 are situated at the 5′ end on the erythromycin fragment with erm 5.1 reading out closest to the end.
15. Many of the lanes on the gel often contain several DNA bands. However, after a PCR cleanup procedure, the samples, including the ones harboring several DNA bands, can be sequenced using the erm primer from the second PCR. The same band pattern on the gel from different colonies, but from the same transposition, indicates that the mutants are siblings. If no product after PCR is obtained, try with a different batch of the arbitrary primer (here termed arb 2).

Acknowledgments

This work was supported by grants from the Danish Council for Independent Research and the Lundbeck Foundation.

References

1. Choi KH, Kim KJ (2009) Applications of transposon-based gene delivery system in bacteria. J Microbiol Biotechnol 19(3):217–228
2. Spatafora G, Rohrer K, Barnard D, Michalek S (1995) A *Streptococcus mutans* mutant that synthesizes elevated levels of intracellular polysaccharide is hypercariogenic in vivo. Infect Immun 63(7):2556–2563
3. Gutierrez JA, Crowley PJ, Brown DP, Hillman JD, Youngman P, Bleiweis AS (1996) Insertional mutagenesis and recovery of interrupted genes of *Streptococcus mutans* by using transposon Tn917: preliminary characterization of mutants displaying acid sensitivity and nutritional requirements. J Bacteriol 178 (14):4166–4175
4. Boyd DA, Cvitkovitch DG, Bleiweis AS, Kiriukhin MY, Debabov DV, Neuhaus FC, Hamilton IR (2000) Defects in D-alanyl-lipoteichoic acid synthesis in *Streptococcus mutans* results in acid sensitivity. J Bacteriol 182 (21):6055–6065
5. Slater JD, Allen AG, May JP, Bolitho S, Lindsay H, Maskell DJ (2003) Mutagenesis of *Streptococcus equi* and *Streptococcus suis* by transposon Tn917. Vet Microbiol 93 (3):197–206
6. Thibessard A, Fernandez A, Gintz B, Decaris B, Leblond-Bourget N (2002) Transposition of pGh9:ISS1 is random and efficient in *Streptococcus thermophilus* CNRZ368. Can J Microbiol 48(5):473–478
7. Lampe DJ (2010) Bacterial genetic methods to explore the biology of mariner transposons. Genetica 138(5):499–508
8. Lampe DJ, Akerley BJ, Rubin EJ, Mekalanos JJ, Robertson HM (1999) Hyperactive transposase mutants of the Himar1 mariner transposon. Proc Natl Acad Sci U S A 96 (20):11428–11433
9. Picardeau M (2010) Transposition of fly mariner elements into bacteria as a genetic tool for mutagenesis. Genetica 138(5):551–558
10. Lampe DJ, Churchill ME, Robertson HM (1996) A purified mariner transposase is sufficient to mediate transposition in vitro. EMBO J 15(19):5470–5479
11. Li M, Rigby K, Lai Y, Nair V, Peschel A, Schittek B, Otto M (2009) *Staphylococcus aureus* mutant screen reveals interaction of the human antimicrobial peptide dermcidin with membrane phospholipids. Antimicrob Agents Chemother 53(10):4200–4210
12. Nilsson M, Christiansen N, Hoiby N, Twetman S, Givskov M, Tolker-Nielsen T (2014) A mariner transposon vector adapted for mutagenesis in oral streptococci. MicrobiologyOpen 3(3):333–340
13. Nilsson M, Rybtke M, Givskov M, Hoiby N, Twetman S, Tolker-Nielsen T (2016) The dlt genes play a role in antimicrobial tolerance of *Streptococcus mutans* biofilms. Int J Antimicrob Agents 48(3):298–304
14. Biswas I, Drake L, Johnson S, Thielen D (2007) Unmarked gene modification in *Streptococcus mutans* by a cotransformation strategy with a thermosensitive plasmid. BioTechniques 42(4):487–490

Chapter 5

Implementation of Transposon Mutagenesis in *Bifidobacterium*

Lorena Ruiz and Douwe van Sinderen

Abstract

Random transposon mutagenesis allows for relatively rapid, genome-wide surveys to detect genes involved in functional traits, by performing screens of mutant libraries. This approach has been widely applied to identify genes responsible for activities of interest in multiple eukaryote and prokaryote organisms, although most studies on microorganisms have focused on pathogenic and clinically relevant bacteria. In this chapter we describe the implementation of an in vitro Tn5-based transposome strategy to generate a large collection of random mutants in the gut commensal *Bifidobacterium breve* UCC2003, and discuss considerations when applying this mutagenesis system to other *Bifidobacterium* species or strains of interest.

Key words Tn5, Transposon, *Bifidobacterium*

1 Introduction

Bifidobacteria are Gram-positive commensal microorganisms whose presence in the human gastrointestinal tract has been associated with beneficial effects on host health [1]. Consequently, significant scientific and commercial efforts have been made to discern the mechanisms responsible of their purported beneficial attributes and cross-talk with the human host and with other members of the intestinal microbiota [2]. In this context, gene inactivation may be considered to represent the gold standard methodology to unequivocally prove the function of specific genes although most *Bifidobacterium* strains remain recalcitrant to genetic manipulation with only a few species and strains having been successfully mutated to date [3–6]. An increasing number of genome-based studies have allowed for insights into the reasons why genetic manipulation of this group of commensal microorganisms is so difficult, and has provided new opportunities to design tailor-made tools to achieve gene inactivation in various *Bifidobacterium* species or strains [4, 5, 7–10].

Transposon-based random mutagenesis approaches generally involve the in vivo delivery of a conditional vector carrying a

Steven C. Ricke et al. (eds.), *Microbial Transposon Mutagenesis: Protocols and Applications*, Methods in Molecular Biology, vol. 2016, https://doi.org/10.1007/978-1-4939-9570-7_5, © Springer Science+Business Media, LLC, part of Springer Nature 2019

transposon piece of DNA containing a selective marker (usually an antibiotic resistance cassette) and the transposase encoding gene. Tight control of transposase expression allows for the transposon to jump into (random positions in) the targeted genome, thereby inactivating any gene in which it had been inserted. Subsequent plasmid curing, which removes the transposase-encoding gene from the cells, fixes the transposon in the original insertion position of the host genome [11]. This approach allows for the generation of an unlimited number of transposon insertion mutants from a single transformant carrying the delivery vector. However, substantial limitations exist to apply such an approach in bifidobacteria as conditional replicative vectors are currently not available for most species or strains. Besides, most heterologous expression systems in bifidobacteria have made use of strong constitutive promoters, which are not appropriate to control the expression of a transposase as high expression of a transposase in the presence of the corresponding transposon, may destabilize the transposon position and ultimately affect cell viability. An alternative to this in vivo transposon delivery approach was proposed early in 2002 when certain mobile genetic elements, that is, Tn5 and Mu transposons, were shown to form functional transposon–transposase complexes by means of in vitro assembly. These complexes were reported to be stable enough to be directly electroporated into the targeted host, where they successfully transposed into the host DNA [12, 13]. The practical implementation of this system in new strains requires achieving (1) high transformation efficiencies in the selected host and (2) the use of an appropriate antibiotic selection marker within the transposon as it needs to be efficiently expressed to confer antibiotic resistance so as to allow for selection of (single-copy) transposon insertion events. The development of these in vitro transposon delivery approaches represented a huge advancement as they offered new opportunities to tackle transposon-based random mutagenesis in microorganisms, such as bifidobacteria.

2 Materials

2.1 Materials and Equipment for Molecular Biology Techniques Used for Transposon Construction, Purification and Assembly; and for Identifying Transposon Insertion Points in the Mutants Obtained

1. Standard molecular biology reagents (*see* **Note 1**): PCR master mix; appropriate restriction enzymes for transposon end pruning (PshAI or PvuII if using pMOD2-TetW), and, if relevant, for cloning a new antibiotic resistance cassette in between the Tn5 mosaic ends to create a new transposon (*see* **Note 2**); plasmid mini-preparation kits to conduct plasmid extraction from bacterial cultures; PCR purification kits; agarose, 1× TAE buffer (40 mM Tris–HCl pH 7.6, 20 mM acetic acid, 1 mM EDTA), ethidium bromide.
2. Replicative vector containing an antibiotic resistant Tn5 transposon (e.g., pMOD2-TetW from Ruiz et al. [14]) or, alternatively, another customized transposon (*see* **Note 2**). This vector

will be used as a source of the transposon piece of DNA and can be stably maintained into appropriate host cells as it does not contain the transposase-encoding gene (and therefore will not destabilize the transposon piece of DNA). If using pMOD2-TetW, most *E. coli* cloning hosts are capable to support its replication, although *endA*$^{-}$ strains are recommended in order to achieve high plasmid yields (*see* **Note 3**).

3. Oligonucleotides to amplify the Tn5 transposon from the replicative vector into which it is being maintained (e.g., pMOD2-TetW). If using pMOD2-TetW, pMOD<MCS>Fw (*5′-ATT CAGGCTGCGCAACTGT-3′*) and pMOD<MCS>Rev (*5′-G TCAGTGAGCGAGGAAGCGGAAG-3′*) can be used as previously described (Ruiz et al. [14]).
4. Purified EZ-Tn5 transposase (commercially available by Lucigen Corporation, Middleton, WI, US—www.lucigen.com), it is a mutated version of the original Tn5 transposase that displays a transposition frequency which is 1000-fold higher than the one exhibited by the original Tn5 transposase.
5. Molecular grade glycerol.

2.2 Microbiology Reagents for Culturing Bacteria

1. Luria–Bertani (LB) culture media: 10 g l^{-1} tryptone (Merck, Darmstadt, Germany), 5 g l^{-1} yeast extract (Merck), 10 g l^{-1} sodium chloride (Merck), sterilized by autoclaving. This medium will be used for routine growth of *E. coli* cells or, supplemented with 10 μg ml^{-1} tetracycline in order to select for *E. coli* strains harboring pMOD2-TetW.
2. de Man–Rogosa–Sharpe (MRS) adjusted to pH 6.8: Proteose Peptone (10 g l^{-1}) (Difco, Franklin Lakes, NJ, USA), beef extract (10 g l^{-1}) (Difco), yeast extract (5 g l^{-1}) (Difco), polysorbate (Tween 80) (1 ml l^{-1}) (Sigma-Aldrich; St Louis, MO, US), triammonium citrate (2 g l^{-1}) (Sigma-Aldrich), magnesium sulfate heptahydrated (0.525 g l^{-1}) (Merck), manganese sulfate tetrahydrated (0.12 g l^{-1}) (Merck), dipotassium phosphate (3 g l^{-1}) (Merck), potassium phosphate (3 g l^{-1}) (Merck), pyruvic acid (0.2 g l^{-1}) (Sigma-Aldrich), cysteine–HCl (0.3 g l^{-1}) (Sigma-Aldrich), ferric sulfate heptahydrate (0.034 g l^{-1}) (Sigma-Aldrich), sterilized by autoclaving. This medium will be used to grow the appropriate *Bifidobacterium* strain for competent cell preparation prior to electroporation.
3. 10% D-lactose solution prepared in distilled water and sterilized by filtration (0.2 μm diameter pore) (*see* **Note 4**).
4. 10% L-cysteine–HCl solution prepared in distilled water and sterilized by filtration (0.2 μm diameter pore).
5. Washing buffer for preparation of electrocompetent cells of *Bifidobacterium:* 1 mM citrate, 0.5 M sucrose buffer, pH 5.8, sterilized by autoclaving.

6. Reinforced Clostridial Medium (RCM) (Oxoid) for routine growth of *Bifidobacterium* cells, resuscitation and recovery of electrotransformed cells.
7. Reinforced Clostridial Agar (RCA) (Oxoid) plates supplemented with appropriate antibiotic concentrations for transposon insertion selection. In the particular case of using the Tn5-TetW transposon described for *B. breve* UCC2003 or *B. breve* NCFB2258 strains [14], use a final concentration of 10 μg ml^{-1} of tetracycline (*see* **Note 5**).
8. 10 mg ml^{-1} tetracycline stock solution dissolved in 50% ethanol and sterilized by filtration (0.2 μm diameter pore). The filtered-sterilized stock solution can be stored at −20 °C (*see* **Note 6**).
9. Plasmid capable of replicating into the selected *Bifidobacterium* host, preferably harboring the same antibiotic resistance cassette than the transposon, to be used as a positive control to test transformation efficiency of the prepared competent cells. If using pMOD2-TetW, the *E. coli–Bifidobacterium* shuttle vector pAM5 [15] can be used as a control plasmid.

2.3 Other Equipment Required

1. Thermocycler.
2. Microcentrifuge.
3. Electroporator and electroporation cuvettes (2 mm gap).
4. Device for spectrophotometric DNA quantification (e.g., NanoDrop or Qubit systems).
5. Electrophoresis unit and transilluminator for DNA visualization.
6. Rotatory shaker incubator.
7. Anaerobic work station (10% H_2, 10% CO_2, 80% N_2).
8. Refrigerated centrifuge with rotor for bottles of at least 50 ml volume.

3 Methods

A schematic overview of the main steps which are detailed in the following sections is represented in Fig. 1.

3.1 Transposon Construction and Preparation

1. A customized transposon can be constructed by cloning the desired antibiotic resistance cassette, or any alternative selective marker to identify transposon insertion events, in between the Tn5 mosaic ends recognized by the Tn5 transposase (Johnson and Reznikoff [16]) in an *E. coli* replicative vector (e.g., pMOD2 vector from Lucigen) (Fig. 1). The *tetW* resistance cassette used in pMOD2-TetW was originally isolated from a

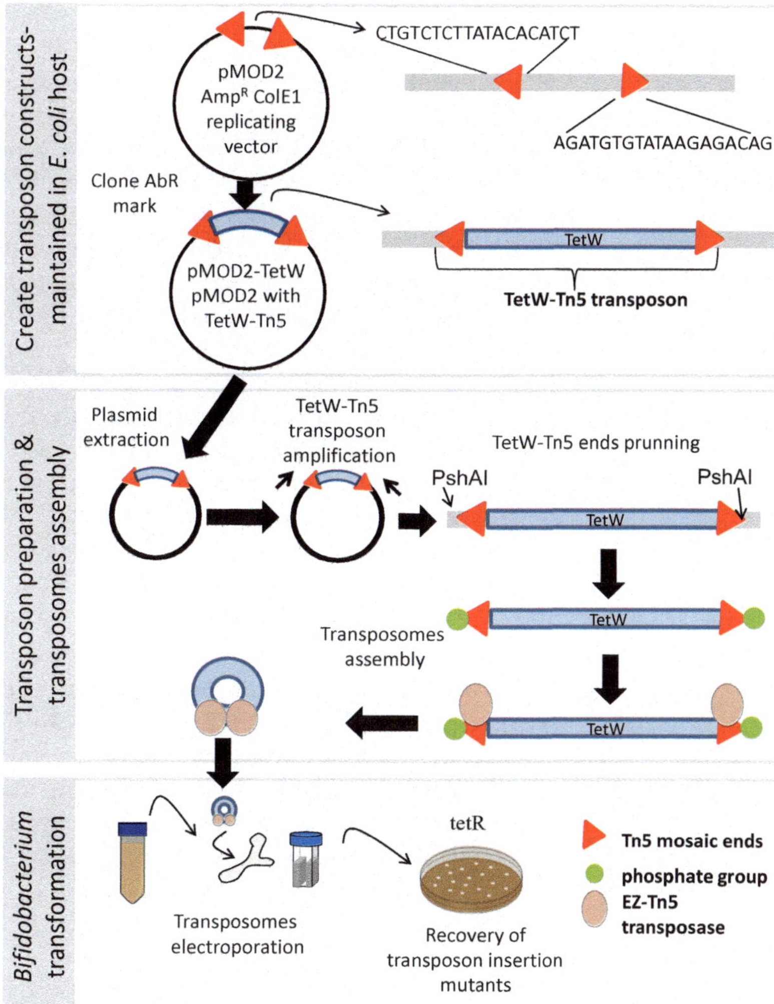

Fig. 1 Schematic overview of the main steps required to create transposon insertion mutants in bifidobacteria, by using EZ-tn5 transposome complexes and a customized tetracycline-resistance Tn5 transposon encompassing a TetW cassette

Bifidobacterium species [17] and has been shown to be efficiently expressed to provide sufficient tetracycline resistance in order to achieve selection of clones harboring a single chromosomal copy of the corresponding gene in multiple *Bifidobacterium* strains [4, 9]. For the above reasons, this is a good marker to be included in a transposon to be used in tetracycline-sensitive *Bifidobacterium* strains [14].

2. Plasmid mini-preparations from the transposon containing construct (e.g., pMOD2-TetW or alternative customized constructs as indicated in **Note 2**) will be used to generate large quantities of a TetW-Tn5 transposon ready for assemblage with

the purified EZ-Tn5 transposase. Plasmid miniprep extractions need to be performed on *E. coli* cells harboring the transposon-containing plasmid (e.g., pMOD2-TetW). An appropriate volume of growing cells must be used according to guidelines of the plasmid mini-preparation kit provider. Extracted plasmid DNA must be verified by restriction profiling followed by agarose gel electrophoresis. For instance, to verify the pMOD2-TetW construct, digest 10 μl of a plasmid mini-preparation in a final reaction volume of 20 μl by using SphI and XbaI according to indications from the restriction enzymes provider. Following incubation for at least 1 h at 37 °C, mix 10 μl of restricted plasmid DNA with 2 μl of commercial loading buffer dye and load on a 1% agarose gel. Apply a voltage of 1.5 V/cm until the dye reaches about two-thirds of the gel length; and stain the gel for 30 min with ethidium bromide (0.5 μg ml^{-1}). Visualize DNA bands in a UV transilluminator. SphI- and XbaI-digested pMOD2-TetW should be visible as bright bands of 2.5 Kb (pMOD2 backbone) and 2.8 Kb (tetracycline-resistant Tn5 transposon).

3. PCR-amplify the transposon piece of DNA using as a template the plasmid mini-preparation of the construct harboring it (e.g., pMOD2-TetW, isolated from *E. coli* cells in the previous step). Oligonucleotides annealing immediately upstream of the inverted repeats flanking the Tn5 transposon in the plasmid harboring the transposon must be used. If using a transposon constructed within a pMOD2 vector, as the one described above for bifidobacteria, oligonucleotides pMOD<MCS>Fw and pMOD<MCS>Rev (Lucigen) can be used. A suggested PCR cycling scheme is as follows: (1) denature the template at 94 °C for 2 min; (2) perform 30 cycles denaturing at 94 °C for 30 s; annealing at 60 °C for 45 s and extending at 72 °C for 1 min for every kb of expected product. Since a good amount of transposon DNA needs to be generated, it is recommended to perform at least 5–10 PCR reactions in a final volume of 50 μl each to guarantee a sufficiently large quantity of transposon DNA (as a general guideline, it would be recommended to generate at least 50–60 μg of PCR product).

4. To verify that PCR reactions have amplified a fragment of the desired size, load 5 μl of the PCR reactions in a 1% agarose gel as previously described in **step 2**. The TetW-Tn5 transposon from Ruiz et al. [14] should produce a band of about 2.8 kb. Each PCR product must be purified on a silica column (*see* **Note 7**).

5. The efficiency of the transposition process is maximized in the presence of phosphorylated transposon ends. In order to get phosphorylated ends, the TetW-Tn5 transposon amplified from pMOD2-TetW as described in the previous step, can be

pruned by restriction with PshAI or PvuII, as these restriction sites are located immediately upstream of the transposon inverted repeats. The restriction reactions must be set up according to guidelines issued by the restriction enzyme provider (*see* **Note 8**). Following restriction, the transposon is cleaned up using a column PCR purification system and each DNA preparation is concentrated by eluting it in a small volume in the final step of the cleaning procedure (20 μl per purification column).

6. Pool all transposon preparations and measure DNA concentration for instance using spectrophotometric methods (e.g., NanoDrop or Qubit-based nucleic acid quantification). The concentration of transposon ends need to be adjusted to desired values taking into consideration the transposon length, by following instructions from the Tn5 transposase provider. The TetW-Tn5 transposon previously used for bifidobacteria was adjusted to a final concentration of 400 ng μl^{-1}.
7. The transposon-transposase complexes are assembled according to instructions from the transposase provider. As a standard guideline, mix 2 μl of transposon DNA preparation (dissolved in TE Buffer (10 mM Tris–HCl (pH 7.5), 1 mM EDTA)), 4 μl of EZ-Tn5 transposase (1 U/μl), and 2 μl of 100% glycerol, vortex, and incubate at room temperature for 30 min. This preparation can be directly electroporated into freshly prepared *Bifidobacterium* competent cells, or stored at −20 °C until use.
8. Assembled transposome complexes can then be electroporated into freshly prepared *Bifidobacterium* electrocompetent cells by following the instructions provided in the following section.

3.2 Preparation of Electrocompetent Cells of Bifidobacterium

In order to achieve maximum transformation efficiencies, electroporation needs to be conducted employing freshly prepared *Bifidobacterium* competent cells. Growth media, wash buffer and antibiotic containing plates are recommended to be prepared freshly the day before the experiment starts.

Day 1:

1. An isolated colony of the strain to be mutated is inoculated into 10 ml of RCM broth and grown overnight (~16 h) at 37 °C in standing tubes into an anaerobic chamber (10% CO_2, 10% H_2, 80% N_2 atmosphere).
2. 50 ml tubes containing de Man–Rogosa–Sharpe broth supplemented with a final concentration of 0.05% L-cysteine (freshly added from a 10% filter-sterilized stock prepared in distilled water); and 1% of an appropriate carbon source (for *B. breve* UCC2003 or *B. breve* NCFB2258, use D-lactose freshly added from a 10% filter-sterilized stock solution prepared in distilled

water) can be prereduced and stored overnight in the same anaerobic chamber.

Day 2:

1. The prereduced and prewarmed de Man–Rogosa–Sharpe broth supplemented with the carbon source and L-cysteine, is inoculated (2% v/v) with the culture grown overnight. Incubate the standing tubes at 37 °C in the anaerobic chamber until the cultures reach an $OD_{600\ nm}$ of about 0.5–0.7.
2. While the culture(s) for competent cell preparation are growing, the citrate–sucrose buffer must be placed in an ice box, and an appropriate centrifuge, 1.5 ml Eppendorf tubes, pipettes, and electroporation cuvettes must be refrigerated by keeping them in an ice box.
3. Incubate cultures for competent cells preparation until they reach an optical density at 600 nm between 0.5 and 0.7 (5–7 h). At this point, take the cultures out of the anaerobic incubator and spin down the cells at 4000 g in a cold rotor (4 °C) for 10 min. *From this point onward the cells need to be kept ice-cold.*
4. Wash the cells twice by decanting the supernatant and gently swirling the cell pellet into the same volume (50 ml) of ice-cold 1 mM citrate–0.5 M sucrose buffer.
5. Resuspend the washed cell pellet obtained from 50 ml of culture into 500 μl of ice-cold citrate–sucrose buffer.
6. In 1.5 ml Eppendorf tubes dispense adequate amounts of Tn5 transposase–transposon complexes assembled as previously described in Subheading 1. In parallel, in one of the tubes add 5 μl of a plasmid capable of replicating into the *Bifidobacterium* strain used (e.g., pAM5 [15]), in order to be used as a positive control to verify the quality of the prepared competent cells and *Bifidobacterium* electrotransformation procedure.
7. Add to each tube containing the transposase–transposon complexes or positive control plasmid DNA, 45 μl of freshly prepared electrocompetent cells. Mix by gently pipetting up and down and transfer the whole volume to an ice-cold electroporation cuvette (*see* **Note 9**).
8. Apply a pulse of 2.0 KV, 200 Ω, 25 μF in an electroporator (e.g., Gene Pulser II Porator Electroporation System from Bio-Rad). Immediately following application of the electric pulse, aseptically add 950 μl of RCM and gently suspend the cells by pipetting up and down a few times.
9. Incubate the electroporation cuvettes with the cell suspensions for 30 min at 37 °C in the anaerobic incubator (*see* **Note 10**).

10. Plate 100 μl aliquots of electroporated cells onto freshly prepared plates of RCA supplemented with 10 μg ml^{-1} of tetracycline added from a filter-sterilized 1000× stock solution as previously described (Subheading 2.2).
11. Incubate the plates anaerobically at 37 °C for 2–3 days.

3.3 Recovery of Tetracycline-Resistant Colonies and Identification of Transposon Insertion Sites

If no antibiotic resistant colonies are grown following electroporation of transposome complexes, verify the presence of antibiotic resistant colonies in the reaction performed with the positive control plasmid DNA. If transformation efficiency of this control plasmid is under 10^4 cfu μg^{-1}, repeat the experiment or consider optimizing the competent cells preparation or transformation procedure (*see* **Note 11**).

If (sufficient) antibiotic resistant colonies are grown following electroporation of transposome complexes:

1. Using a sterile pipette tip, inoculate them into RCM broth supplemented with tetracycline (10 μg ml^{-1}) and cultivate them overnight anaerobically at 37 °C.
2. The following day, add 30% sterile glycerol and stock the clones at −80 °C for further analysis.
3. The transposon presence in obtained clones can be verified through (a) Southern blot hybridization, and (b) PCR using oligonucleotides targeting the transposon mosaic ends or targeting internal sequence fragments within the transposon.
4. Identification of transposon insertion sites in specific clones can be performed through inverse-PCR coupled to Sanger sequencing as previously described [14] (*see* **Note 12**). As a general guideline, the inverse-PCR procedure to identify the genome positions where a transposon is inserted in a specific clone includes the following steps: (1) extract *Bifidobacterium* DNA (*see* **Note 13**); (2) digest 1 μg of DNA in a final volume of 50 μl of reaction, using a restriction enzyme which does not cut within the transposon; (3) purify digested DNA; a suggested protocol is as follows: add 50 μl of distilled water and 100 μl of phenol–chloroform pH 8.0, mix and spin down (≥12,000 g, 10 min), transfer upper phase to a fresh tube, add 1/10 volumes of sodium acetate 3 M pH 5.8 and one volume of cold ethanol; spin down 30 min, ≥12,000 g, 4 °C; remove supernatant; wash with 70% ethanol; air-dry and resuspend in 10 μl of distilled water; (4) quantify DNA spectrophotometrically; (5) set up self-ligation reactions, using 0.2 or 0.4 μg of restricted DNA in a final volume of 50 μl; (6) incubate ligations overnight at room temperature; (7) purify ligated DNA using the same procedure described in **step 3**; (8) set up PCR reactions using each ligation reaction as template DNA and oligonucleotides annealing with the transposon ends, outward

facing (If using the TetW-Tn5 previously used for bifidobacteria, the oligonucleotides i-PCR-Fw 5′-GCATACCGTACT GATCTG-3′ and i-PCR-Rev 5′-CAATCATACCGGCTTCC-3′ can be used.) (*see* **Note 14**); (9) verify the PCR amplification by loading 5 μl in a 1% agarose gel as previously described (Subheading 3.1, **step 2**); (10) sequence the PCR products using nested oligonucleotides, located within the transposon ends, upstream of the position where the inverse-PCR oligonucleotides annealed; for example, if using the TetW-Tn5 transposon previously described for bifidobacteria, sequencing can be performed using oligonucleotides pMOD-fw-seq *5′-GCCAACGACTACGCACTAGCC-3′* and pMOD-rev-seq *5′-GAGCCAATATGCGAGAACACC-3′* [14].

4 Notes

1. Molecular biology grade reagents need to be used for transposon construction and transposase–transposon assembly. Standard caution to prevent nuclease contamination of transposon DNA preparation needs to be taken (use molecular biology grade reagents, gloves, and filter tips).
2. Virtually any piece of DNA can be included into a transposon, in between the mosaic ends recognized by the corresponding transposase. For further instructions and background information please check Lucigen guidelines.
3. If using pMOD2-TetW grow the *E. coli* host cells harboring the vector in LB supplemented with 10 μg ml^{-1} of tetracycline. Grow *E. coli* cultures into sterile tubes, ensuring to leave a sufficient empty space in the tubes to allow for appropriate aeration during incubation. Grow the cultures by incubating them overnight at 37 °C in a rotary shaker (200 rpm).
4. D-Lactose might need to be replaced by an alternative carbon source depending on the particular needs of the strain used [14]. To facilitate preparation of a 10% stock solution of D-lactose, the suspension might need to be warmed up at 37 °C (e.g., incubate it in a water bath at 37 °C for 5 min and then vortex until the carbohydrate is fully dissolved in the suspension). Then filter-sterilize (0.2 μm pore filters).
5. After autoclaving agar containing media, leave the bottles to cool down to approximately 50 °C before adding the antibiotic solution as it might be heat labile. Gently mix the medium bottles and pour the plates.
6. Discard the tetracycline solution stored at −20 °C if precipitation is observed.
7. Alternative methods to purify the PCR-amplified transposon so as to eliminate salts, and excess of nucleotides and primers, can

be used (e.g., polyethylene glycol precipitation or standard ethanol precipitation).

8. When designing a new customized transposon construct, ensure the absence of PshAI or PvuII internal restriction sites in the constructed transposon. Alternatively, whenever these two sites have target sequences within the transposon sequence, it is possible to PCR-amplify the transposon piece of DNA by employing a single phosphorylated oligonucleotide, annealing with the Tn5 mosaic end (*5′-CTGTCTCTTA TACACATCT-3′*). If the transposon is amplified in this way from the pMOD-derived constructs, there is no need to trim the transposon ends with PshAI or PvuII restriction.
9. Avoid the presence of bubbles in the sample dispensed in the electroporation cuvette.
10. Time of incubation following electroporation might need to be adjusted depending on the strains used and transformation efficiencies. Longer times increase the number of antibiotic resistant colonies obtained, but this may be the result of growth rather than increased transposition efficiency.
11. When optimizing transposon mutagenesis strategies in new strains, it is critical to achieve high transformation efficiencies and to guarantee that the antibiotic selective marker included within the transposon is efficiently expressed in the selected host and allows for accurate and reliable selection for transposition events. Type I and Type II endogenous restriction-modification systems encoded in the host genome have been demonstrated as key bottlenecks limiting transformation efficiencies in bifidobacteria [4, 9]. Therefore, if the (recognition sites of the) restriction-modification systems of the host genome are known, it is advisable to avoid the presence of the corresponding restriction sites in the transposon sequence.
12. For alternative procedures to simultaneously identify transposon insertion points in large collections of transposon insertion mutants, *see* [18].
13. Multiple procedures have been available in order to extract DNA from *Bifidobacterium* cells. As a general guideline, cell pellets obtained from 6 ml of *Bifidobacterium* cells grown overnight need to be lysed enzymatically by suspending the cell pellets into TE buffer supplemented with lysozyme (30 mg ml^{-1}) and mutanolysin (5 U ml^{-1}) and incubated for at least 30 min at 37 °C. Then proceed with DNA purification procedures either using column-based kits or standard phenol–chloroform purification coupled to ethanol precipitation [19].
14. In the inverse-PCR program, include a long elongation time in each cycle (at least 8 min).

References

1. Tojo R, Suárez A, Clemente MG et al (2014) Intestinal microbiota in health and disease: role of bifidobacteria in gut homeostasis. World J Gastroenterol 20:15163–15176. https://doi.org/10.3748/wjg.v20.i41.15163
2. Bottacini F, van Sinderen D, Ventura M (2017) Omics of bifidobacteria: research and insights into their health-promoting activities. Biochem J 474:4137–4152. https://doi.org/10.1042/BCJ20160756
3. Brancaccio VF, Zhurina DS, Riedel CU (2013) Tough nuts to crack: site-directed mutagenesis of bifidobacteria remains a challenge. Bioengineered 4:197–202. https://doi.org/10.4161/bioe.23381
4. O'Callaghan A, Bottacini F, O'Connell Motherway M, van Sinderen D (2015) Pan-genome analysis of *Bifidobacterium longum* and site-directed mutagenesis through by-pass of restriction-modification systems. BMC Genomics 16:832. https://doi.org/10.1186/s12864-015-1968-4
5. Hidalgo-Cantabrana C, Sánchez B, Álvarez-Martín P et al (2015) A single mutation in the gene responsible for the mucoid phenotype of *Bifidobacterium animalis* subsp. *lactis* confers surface and functional characteristics. Appl Environ Microbiol 81:7960–7968. https://doi.org/10.1128/AEM.02095-15
6. Sakaguchi K, He J, Tani S et al (2012) A targeted gene knockout method using a newly constructed temperature-sensitive plasmid mediated homologous recombination in *Bifidobacterium longum*. Appl Microbiol Biotechnol 95:499–509. https://doi.org/10.1007/s00253-012-4090-4
7. O' Connell Motherway M, Watson D, Bottacini F et al (2014) Identification of restriction-modification systems of *Bifidobacterium animalis* subsp. *lactis* CNCM I-2494 by SMRT sequencing and associated methylome analysis. PLoS One 9:e94875. https://doi.org/10.1371/journal.pone.0094875
8. Hirayama Y, Sakanaka M, Fukuma H et al (2012) Development of a double-crossover markerless gene deletion system in *Bifidobacterium longum*: functional analysis of the α-galactosidase gene for raffinose assimilation. Appl Environ Microbiol 78(14):4984–4994. https://doi.org/10.1128/AEM.00588-12
9. O'Connell Motherway M, O'Driscoll J, Fitzgerald GF, Van Sinderen D (2009) Overcoming the restriction barrier to plasmid transformation and targeted mutagenesis in *Bifidobacterium breve* UCC2003. Microb Biotechnol 2:321–332. https://doi.org/10.1111/j.1751-7915.2008.00071.x
10. Bottacini F, Morrissey R, Roberts RJ et al (2018) Comparative genome and methylome analysis reveals restriction/modification system diversity in the gut commensal *Bifidobacterium breve*. Nucleic Acids Res 46:1860–1877. https://doi.org/10.1093/nar/gkx1289
11. Picardeau M (2010) Transposition of fly mariner elements into bacteria as a genetic tool for mutagenesis. Genetica 138:551–558. https://doi.org/10.1007/s10709-009-9408-5
12. Reznikoff WS, Goryshin IY, Jendrisak JJ (2004) Tn5 as a molecular genetics tool: in vitro transposition and the coupling of in vitro technologies with in vivo transposition. Methods Mol Biol 260:83–96. https://doi.org/10.1385/1-59259-755-6:083
13. Lamberg A, Nieminen S, Qiao M, Savilahti H (2002) Efficient insertion mutagenesis strategy for bacterial genomes involving electroporation of in vitro-assembled DNA transposition complexes of bacteriophage mu. Appl Environ Microbiol 68:705–712. https://doi.org/10.1128/AEM.68.2.705-712.2002
14. Ruiz L, Motherway MO, Lanigan N, van Sinderen D (2013) Transposon mutagenesis in *Bifidobacterium breve*: construction and characterization of a Tn5 transposon mutant library for *Bifidobacterium breve* UCC2003. PLoS One 8:e64699. https://doi.org/10.1371/journal.pone.0064699
15. Alvarez-Martin P, Belen Florez A, Margolles A et al (2008) Improved cloning vectors for bifidobacteria, based on the *Bifidobacterium catenulatum* pBC1 replicon. Appl Environ Microbiol 74:4656–4665. https://doi.org/10.1128/AEM.00074-08
16. Johnson RC, Reznikoff WS (1983) DNA sequences at the ends of transposon Tn5 required for transposition. Nature 304:280–282. https://doi.org/10.1038/304280a0
17. Flórez AB, Ammor MS, Alvarez-Martín P et al (2006) Molecular analysis of tet(W) gene-mediated tetracycline resistance in dominant intestinal *Bifidobacterium* species from healthy humans. Appl Environ Microbiol 72:7377–73799. https://doi.org/10.1128/AEM.00486-06
18. Barquist L, Boinett CJ, Cain AK (2013) Approaches to querying bacterial genomes with transposon-insertion sequencing. RNA Biol 10:1161–1169. https://doi.org/10.4161/rna.24765
19. Sambrook J, Russell DW (2006) Purification of nucleic acids by extraction with phenol:chloroform. CSH Protoc 2006(1). https://doi.org/10.1101/pdb.prot4455

Chapter 6

Transposon Mutagenesis of *Listeria monocytogenes*

Oindrila Paul, Damayanti Chakravarty, and Janet R. Donaldson

Abstract

Listeria monocytogenes is a Gram-positive, facultative intracellular foodborne pathogen that enters the human digestive tract after the consumption of contaminated food. Much research has been done to understand the virulence factors of *Listeria monocytogenes*. One useful tool to study these virulence factors has been transposon mutagenesis. Many mutants can be generated at a time by performing high-throughput mutagenesis using transposons and later screening these mutants to identify features related to particular functions in the bacteria. Many transposon delivery systems are not ideal for transposon studies in *Listeria monocytogenes*, as the transposon system is too large, has lower transposition efficiency, and a high rate of plasmid retention. Therefore, a new *mariner*-based transposition system has been developed for *Listeria monocytogenes*. This system is an ideal high-throughput transposon mutagenesis as the rate of transposition is high and random, along with very low plasmid retention capacity.

Key words *Listeria monocytogenes*, Mariner, Transposon, Mutagenesis

1 Introduction

Listeria monocytogenes is a Gram-positive, foodborne bacterium that causes one of the deadliest food borne illnesses called listeriosis [1]. This bacterium is facultative intracellular, facultative anaerobe, and able to survive a variety of temperatures and environmental stressors [2]. The pathogen primarily affects the elderly, the young, the immunocompromised, and pregnant women. This pathogen has a plethora of virulence factors that contribute to its pathogenicity. To study these factors, transposon mutagenesis can be used to construct random mutants for functional analyses [3]. In fact, transposon mutagenesis has been used in several studies with *Listeria monocytogenes* and has led to many discoveries of novel genes involved in different processes of the bacterium [4–8].

In 1986, Gaillard et al. used the conjugative transposon Tn*1545* to study the role of hemolysin in pathogenesis of *L. monocytogenes* [9]. Tn*1545* is a 26 kb molecule encoding resistance to kanamycin, tetracycline and erythromycin [10]. A single copy of Tn*1545* was inserted into the chromosome of *L. monocytogenes* in the hemolysin

Steven C. Ricke et al. (eds.), *Microbial Transposon Mutagenesis: Protocols and Applications*, Methods in Molecular Biology, vol. 2016, https://doi.org/10.1007/978-1-4939-9570-7_6, © Springer Science+Business Media, LLC, part of Springer Nature 2019

gene. This study demonstrated the importance of hemolysin in the pathogenesis of *L. monocytogenes*. Additionally, this study also proved for the first time that transposon mutagenesis is feasible in *L. monocytogenes*.

Tn*1545* was also used in a signature-tagged transposon mutagenesis of *L. monocytogenes* using a murine model [11]. Using this method, the authors were able to screen 2000 mutants to identify 18 mutants with reduced virulence in the murine model. These mutants corresponded to cell wall components and proteins involved in various cellular processes, such as recombination, transcription regulation, and metabolism.

Tn*916* has also been utilized in studies with *L. monocytogenes*. Tn*916* is a self-mobilizable tetracycline resistance transposon that was transferred to *L. monocytogenes* by conjugation to construct a hemolysin mutant [12]. This study demonstrated that Tn*916* transforms into the *L. monocytogenes* chromosome with a high frequency, approximately 10^{-8} to 10^{-10}. This enables a convenient screening method for mutants in specific genes [12].

The Tn*3*-like transposon Tn*917* has also been shown to be useful for insertional mutagenesis with *L. monocytogenes* [13]. For proper insertional mutagenesis and characterization of disrupted genes, two new derivatives of Tn*917* were assembled: Tn*917*-LTVl and Tn*917*-LTV3. These could transpose at a significantly elevated frequency, generate transcription *lacZ* fusions after it is inserted in an appropriate orientation into a chromosomal gene, and allow for the rapid cloning in *Escherichia coli* of chromosomal DNA flanking transposon insertions. The transposon derivatives carry CoIEl replication functions, a cluster of polylinker cloning sites and antibiotic resistance genes selectable in *E. coli* (*bla* for Tn*917*-LTVl; *neo* and *bk* for Tn*917*-LTV3), which allows for rapid cloning of DNA flanking insertions. The enhanced transposition frequency is thought to be due to the placement of vector-derived promoters upstream from the Tn*917* transposase gene. In *L. monocytogenes*, Tn*917*-LTV3 transposed at a frequency of 10^{-8} when introduced on a pE194Ts-derived vector and generated at least eight different auxotrophic mutations.

Tn*6188* has also been used in *L. monocytogenes* strains 4423 and 6179 [14]. Tn*6188* transposase genes *tnpABC* are significantly similar to transposases of Tn*554*-like transposons. It encodes *qacH*, a transporter of small multidrug-resistant protein family similar to Smr/EmrE/Qac proteins which have increased tolerance against different disinfectants [15].

The last and most recently studied transposons for a novel signature tagged mutagenesis system are the mariner family of transposons [16]. They are abundant in nature and belong to IS630 family of insertion sequences [17, 18]. Himarl has been well studied for mutagenesis in bacteria [17]. It belongs to Tcl/mariner superfamily of transposable elements. This particular type of transposon has many advantages compared to previously used transposons in *L. monocytogenes.* These transposons do not need species-specific host factors for transposition and only need dinucleotide TA for insertion into the chromosome, which is low in GC-rich *Listeria monocytogenes* [17–19]. Other transposon systems like Tn*917* are more likely to target hotspots unlike mariner transposon PJZ037 [13, 20–22].

Together, these previous studies indicate that transposons have a great advantage for *L. monocytogenes* mutagenesis studies. These studies have generated strong data related to functional analyses of genes that otherwise would not had been identified. In particular, the use of mariner transposon systems provides the greatest flexibility for functional analyses in *L. monocytogenes.*

2 Materials

2.1 Bacterial Strains, Culture Conditions, and Plasmids

The *Listeria monocytogenes* strain 10403S is used in the following methods (**Notes 1–8**). *Listeria monocytogenes* is routinely cultured in brain–heart infusion (BHI) media at 37 °C (**Notes 9–10**). Plasmids to construct the Himar1-based transposon include: pPL2 and pDG780. Enzymes needed for the construction of the Himar1-based transposon include: SacI, XhoI, BamHI, KpnI, SphI, and NdeI.

2.2 Electroporation of L. monocytogenes

Listeria monocytogenes is first made electrocompotent by growing in BHI with 0.5 M Sucrose to OD600 = 0.2. Penicillin G (10 μg/ml) is added and cells are incubated for an additional 2 h. Cells are then harvested by centrifugation at 3000 × *g* for 10 min at 4 °C. Pellets are washed with 0.5 M sucrose in 1 mM HEPES solution (**Note 11**). Following wash, cells are once again pelleted and resuspended in 0.5 M Sucrose in 1 mM HEPES supplemented with 15% glycerol. Cells are frozen at −80 °C until further use. Forty microliters of cells is added to 100 ng of plasmid DNA and electroporated at 25 μF, 12.5 kV/cm (**Notes 12–14**). Immediately, 1 ml of fresh BHI media is then added and cells are incubated at 30 °C as described below (**Note 15**).

3 Methods

The Himar1-based transposon is constructed for use with *L. monocytogenes* by Cao et al. [20].

3.1 Constructing the Mariner-Based Transposon Delivery Vectors

1. The plasmid pMC14 is constructed through the SacI and XhoI digestion of pPL2 and pDG780 to clone kanamycin resistance cassette from pDG780 into pPL2 (Fig. 1).

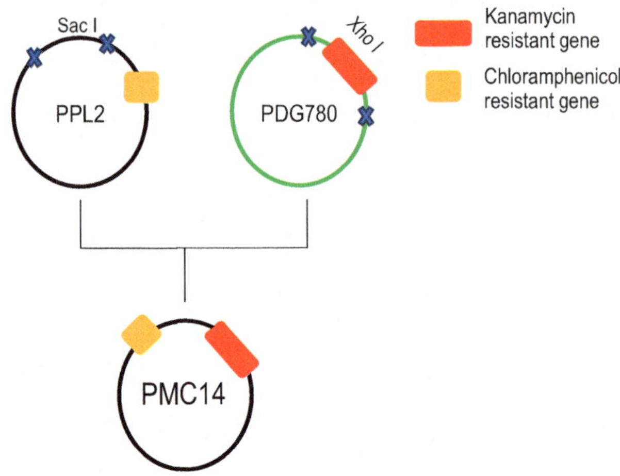

Fig. 1 Construction of PMC14. The plasmid pMC14 is constructed through the SacI and XhoI digestion of pPL2 and pDG780 to clone kanamycin resistance cassette from pDG780 into pPL2

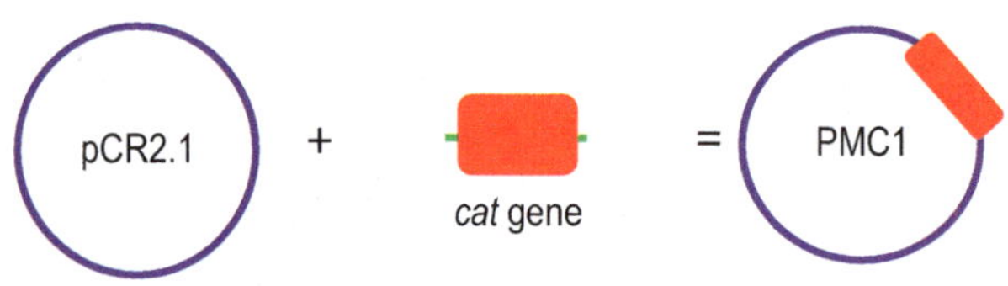

Fig. 2 Generation of pMC1. The primer pair Marq155/156 is used to amplify the chloramphenicol acetyltransferase gene (*cat*) from pPL2 using PCR and is ligated into pCR2.1-Topo (Invitrogen)

2. The primer pair Marq155/156 is used to amplify the chloramphenicol acetyltransferase gene (*cat*) from pPL2 using PCR (5′-TTGGATCCCGGAGACGGTCACA/ CGCATCTGTGCGGTATTTCA-3′).
3. To create pMC1, the amplified product from the previous step is ligated into pCR2.1-Topo (Invitrogen; Fig. 2).
4. BamH1 is used to digest pMC1.
5. The digested product is ligated between the 5′ and 3′ inverted terminal repeats (ITR) of *Himar1* into the BglII site of pMMOrf which contains 5′ and 3′ ITR from *Himar1* to create pMC3 (Fig. 3).
6. The primer pair Marq188/234 is then used to amplify the *Himar1* transposase gene (*tpase*) from pNF1100. This contains a copy of *Himar1* derived from pMEnt-*neo* (5′-ATCCGCATG CTGCAAGGCGATTAAGT-3′/5′-GCGGATCCAG*AGGAG* TTTTATGAATATGGAAAAAAAGGAATTTCGTGTTT-3′).
7. The primer pairs Marq247/248 and Marq249/250 are used to amplify the promoter regions *mrgA* and *katA* genes

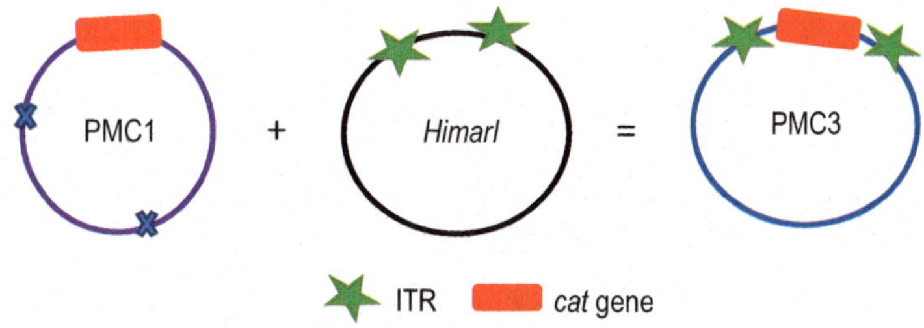

Fig. 3 Generation of pMC3. BamH1 is used to digest pMC1, ligate between the 5′ and 3′ inverted terminal repeats (ITR) of *Himar1* into the BglII site of pMMOrf which contain 5′ and 3′ ITR from *Himar1* to create pMC3

respectively of *Bacillus subtilis* CU1065 (5′-GCGGTACCTATCATCAATACTATA-3′/5′-CAGGATCCGTGATCTGTTGACTTAAT-3′).

8. After digesting the PCR products of P_{mrgA}, P_{katA}, and *tpase* with BamHI, each promoter is ligated to *tpase* individually.
9. The ligation products of P_{mrgA}-*tpase* and P_{katA}-*tpase* are amplified using the primer pairs Marq247/188 and Marq249/188 respectively (5′-GCGGTACCTATCATCAATACTATA-3′/5′-ATCCGCATGCTGCAAGGCGATTAAGT-3′) and (5′-GCGGTACCTTTTCTTTGATGCTGA-3′/5′-ATCCGCATGCTGCAAGGCGATTAAGT-3′).
10. KpnI and SphI are used to digest the PCR products of P_{mrgA}-*tpase* and P_{katA}-*tpase* and pMC14.
11. P_{mrgA}-*tpase* is then ligated with the pMC14 fragment containing the P4*ori*T, p15A*ori* and Gram-negative *cat* whereas P_{katA}-*tpase* is ligated with the pMC14 fragment containing the Gram-positive *kan* genes.
12. From pMC3, the fragment containing ITR-*cat*-ITR is ligated at the KpnI and XhoI sites.
13. *ermC* replaces the *cat* gene from the previous step. This replacement is achieved by amplifying *ermC* from pPL3e using the primer pair Marq205/206. The PCR products and the vector are digested with NdeI and the *cat* gene is replaced with *ermC* between the ITR.
14. The primer pair Marq194/195 is used to amplify the temperature sensitive origin of replication pE194ts *ori* in pKSV7 (5′-CGGGTACCATCACACGCAAAAAGGA-3′/5′-CGGGTACCTAAATTCAAAATCTATC-3′).
15. pMC38 (P_{mrgA}) and pMC39 (P_{katA}) are created by digesting the PCR products with KnpI and ligating into each vector (Fig. 4).

The vectors are 8172 bp for pMC38 and 8297 bp for pMC39 and the transposon is 1395 bp (Fig. 5).

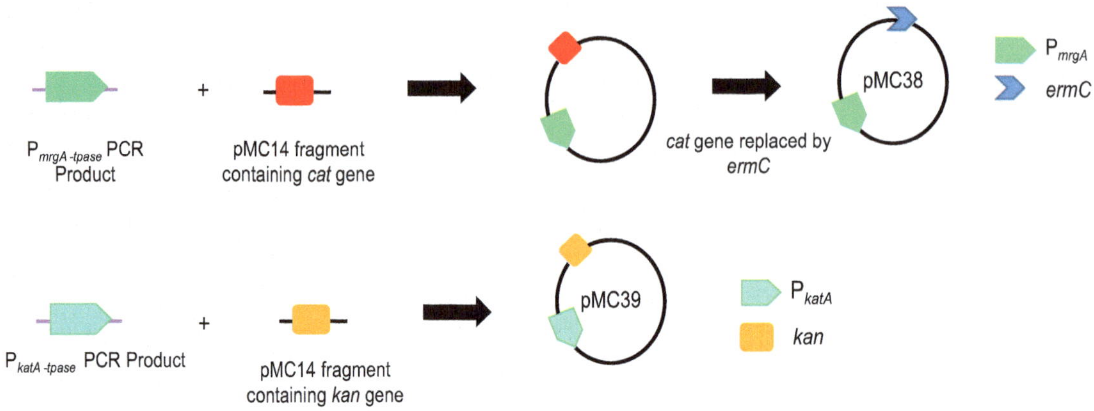

Fig. 4 Generation of pMC38 and pMC39

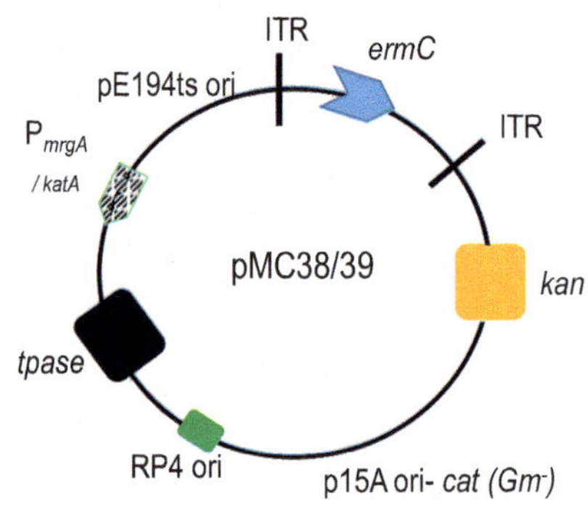

Fig. 5 Physical map of pMC38 and pMCR39. Adapted from [5]. Inverted terminal repeats (ITR) flank the erythromycin cassette

3.2 Evaluating the Mariner*-Based Transposon Delivery*

1. Using electroporation, pMC38 and pMC39 are transformed into *Listeria monocytogenes* 10403S (Fig. 6).
2. Transformants are selected on brain–heart infusion (BHI) plates supplemented with 5 μg/ml erythromycin at 30 °C.
3. Individual colonies are grown at 30 °C with shaking, overnight in BHI supplemented with 10 μg/ml erythromycin and kanamycin.
4. Cultures are grown for an hour at 30 °C with shaking after diluting 1:200 in BHI with erythromycin.
5. After an hour, the cultures are shifted to 40 °C and allowed to grow for 6 h until the optical density at 600 nm reaches between 0.3 and 0.5.
6. Aliquots of culture are plated on BHI supplemented with erythromycin and incubate at 40 °C.

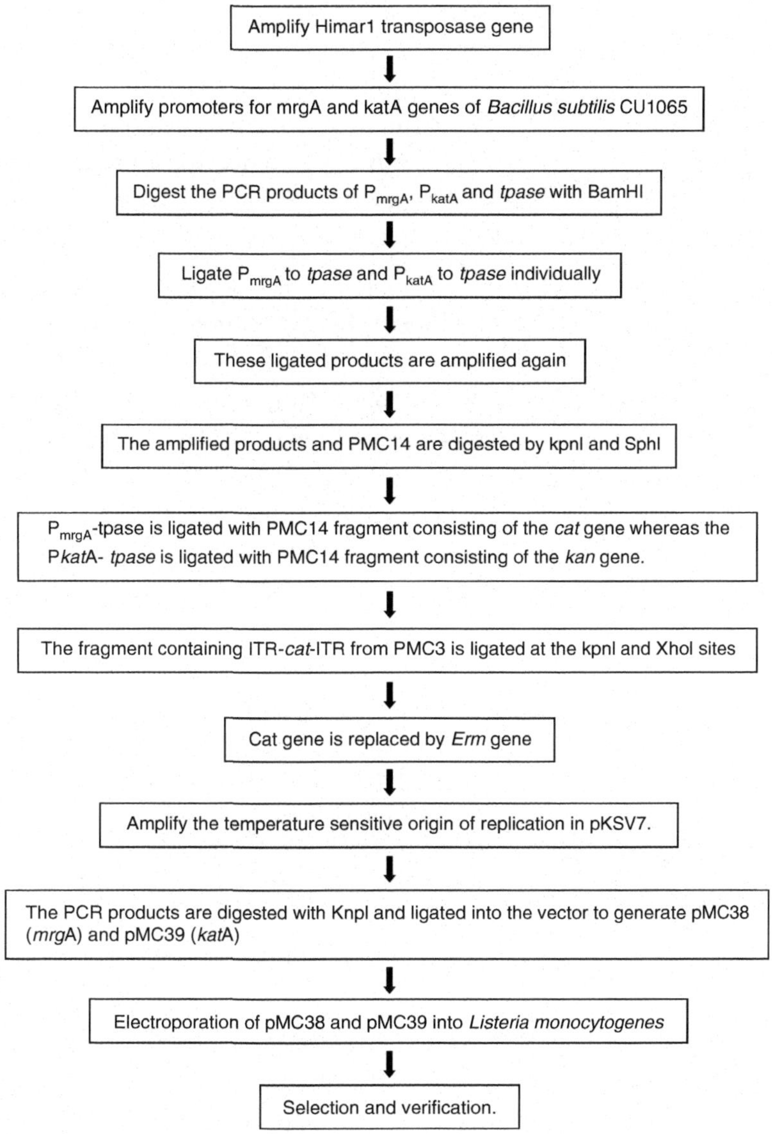

Fig. 6 Schematic of the generation of the mariner transposon

7. To evaluate plasmid retention, individual colonies are picked and plated either on BHI supplemented with erythromycin or kanamycin.

4 Notes

1. *Listeria monocytogenes* is a BSL2 bacterium. Caution must be exercised while handling the live pathogen.
2. Live pathogens should be handled inside the Biological Safety cabinet.

3. Safety goggles must be worn while handling liquid cultures (to protect yourself from splash hazard).
4. Never touch your mouth and eyes wearing gloves.
5. Laboratory handling *Listeria monocytogenes* should have non-porous floor, bench tops, chairs, and stools.
6. Wastes from the experiment should be treated as biohazard and be autoclaved before discarding.
7. Do NOT mouth-pipette.
8. Tips and tubes in molecular biology experiment used should be sterilized (i.e., DNase and RNase free).
9. Add the antibiotic after autoclaving the media.
10. If your plates are in the refrigerator, prewarm before plating the sample.
11. The tubes should be set on ice while making all the transfers during the molecular biology experiments.
12. For electroporation, cuvettes and tubes should be prechilled on ice.
13. Electrocompetent cells should be thawed on ice.
14. If high concentration salt or air bubbles are present in the sample, arcing may occur during electroporation.
15. Add the recovery media to the cells after immediately after electroporation.

References

1. Ryser ET, Marth EH (1989) "New" food-borne pathogens of public health significance. J Am Diet Assoc 89:948–954
2. Farber JM, Peterkin PI (1991) *Listeria monocytogenes*, a food-borne pathogen. Microbiol Rev 55:476–511
3. Kwon YM, Ricke SC, Mandal RK (2016) Transposon sequencing: methods and expanding applications. Appl Microbiol Biotechnol 100:31–43
4. Annous BA, Becker LA, Bayles DO, Labeda DP, Wilkinson BJ (1997) Critical role of anteiso-C15:0 fatty acid in the growth of *Listeria monocytogenes* at low temperatures. Appl Environ Microbiol 63:3887–3894
5. Zhu K, Bayles DO, Xiong A, Jayaswal RK, Wilkinson BJ (2005) Precursor and temperature modulation of fatty acid composition and growth of *Listeria monocytogenes* cold-sensitive mutants with transposon-interrupted branched-chain alpha-keto acid dehydrogenase. Microbiology 151:615–623
6. Angelidis AS, Smith LT, Hoffman LM, Smith GM (2002) Identification of opuC as a chill-activated and osmotically activated carnitine transporter in *Listeria monocytogenes*. Appl Environ Microbiol 68:2644–2650
7. Sleator RD, Wouters J, Gahan CG, Abee T, Hill C (2001) Analysis of the role of OpuC, an osmolyte transport system, in salt tolerance and virulence potential of *Listeria monocytogenes*. Appl Environ Microbiol 67:2692–2698
8. Joseph B, Przybilla K, Stuhler C, Schauer K, Slaghuis J, Fuchs TM, Goebel W (2006) Identification of *Listeria monocytogenes* genes contributing to intracellular replication by expression profiling and mutant screening. J Bacteriol 188:556–568
9. Gaillard JL, Berche P, Sansonetti P (1986) Transposon mutagenesis as a tool to study the role of hemolysin in the virulence of *Listeria monocytogenes*. Infect Immun 52:50–55
10. Collatz E, Carlier C, Courvalin P (1984) Characterization of high-level aminoglycoside

resistance in a strain of *Streptococcus pneumoniae*. J Gen Microbiol 130:1665–1671

11. Autret N, Dubail I, Trieu-Cuot P, Berche P, Charbit A (2001) Identification of new genes involved in the virulence of *Listeria monocytogenes* by signature-tagged transposon mutagenesis. Infect Immun 69:2054–2065
12. Kathariou S, Metz P, Hof H, Goebel W (1987) Tn916-induced mutations in the hemolysin determinant affecting virulence of *Listeria monocytogenes*. J Bacteriol 169:1291–1297
13. Camilli A, Portnoy A, Youngman P (1990) Insertional mutagenesis of *Listeria monocytogenes* with a novel Tn917 derivative that allows direct cloning of DNA flanking transposon insertions. J Bacteriol 172:3738–3744
14. Muller A, Rychli K, Muhterem-Uyar M, Zaiser A, Stessl B, Guinane CM, Cotter PD, Wagner M, Schmitz-Esser S (2013) Tn6188 - a novel transposon in *Listeria monocytogenes* responsible for tolerance to benzalkonium chloride. PLoS One 8:e76835
15. Muller A, Rychli K, Zaiser A, Wieser C, Wagner M, Schmitz-Esser S (2014) The *Listeria monocytogenes* transposon Tn6188 provides increased tolerance to various quaternary ammonium compounds and ethidium bromide. FEMS Microbiol Lett 361:166–173
16. Cummins J, Casey PG, Joyce SA, Gahan CG (2013) A mariner transposon-based signature-tagged mutagenesis system for the analysis of oral infection by *Listeria monocytogenes*. PLoS One 8:e75437. https://doi.org/10.1371/journal.pone.0075437
17. Picardeau M (2010) Transposition of fly mariner elements into bacteria as a genetic tool for mutagenesis. Genetica 138:551–558
18. Plasterk RH, Izsvak Z, Ivics Z (1999) Resident aliens: the Tc1/mariner superfamily of transposable elements. Trends Genet 15:326–332
19. Lampe DJ, Churchill ME, Robertson HM (1996) A purified mariner transposase is sufficient to mediate transposition in vitro. EMBO J 15:5470–5479
20. Cao M, Bitar AP, Marquis H (2007) A mariner-based transposition system for *Listeria monocytogenes*. Appl Environ Microbiol 73:2758–2761
21. Garsin DA, Urbach J, Huguet-Tapia JC, Peters JE, Ausubel FM (2004) Construction of an *Enterococcus faecalis* Tn917-mediated-gene-disruption library offers insight into Tn917 insertion patterns. J Bacteriol 186:7280–7289
22. Zemansky J, Kline BC, Woodward JJ, Leber JH, Marquis H, Portnoy DA (2009) Development of a mariner-based transposon and identification of *Listeria monocytogenes* determinants, including the peptidyl-prolyl isomerase PrsA2, that contribute to its hemolytic phenotype. J Bacteriol 191:3950–3964

Chapter 7

Transposon Mutagenesis of Foodborne Pathogenic *Escherichia coli*

Supraja Puttamreddy and F. Chris Minion

Abstract

The *Enterobacteriaceae*, and in particular, *Escherichia coli* including foodborne pathotypes are particularly amenable to transposon mutagenesis. Here we describe the use of mini-Tn*5* and Mu d1(Ap *lac*) to generate transposon inserts for analysis of enterohemorrhagic *Escherichia coli* EDL933. We also discuss how to array the library in 96-well plates and sequence individual clones for further analysis.

Key words Enterohemorrhagic *Escherichia coli*, Transposons, Library construction, DNA sequence analysis, Mu phage

1 Introduction

Transposon mutagenesis is a powerful tool for the analysis of bacterial features. The ability to construct a library of well-defined mutations in a target species is often the first step in identification and characterization of genes involved in specific phenotypes. For pathogenic *E. coli*, transposons can take a variety of forms, from standard transposons like Tn*5* and Tn*10* to mini-transposons (mini-Tn*5*) where the transposase is outside the inverted repeats to enhance stability of the insert, to constructs that carry different types of reporter genes for producing gene and operon fusions [1] (*see* **Note 1**). Then there are phage that while not technically considered transposons (e.g., Mu), nonetheless replicate by transposition and have the ability to generate the same kind of mutations of reporter fusions (e.g., Mu d1 (Ap *lac*)) if appropriately modified (*see* **Note 2**). The latter will not be further discussed in this chapter, but many of the downstream techniques following mutagenesis are equally applicable to them. With a large selection of transposon variants, the issues are more related to delivery rather than what construct to use.

Steven C. Ricke et al. (eds.), *Microbial Transposon Mutagenesis: Protocols and Applications*, Methods in Molecular Biology, vol. 2016, https://doi.org/10.1007/978-1-4939-9570-7_7, © Springer Science+Business Media, LLC, part of Springer Nature 2019

Efficient delivery systems depend on several common features. First, the transposon must be moved to the recipient cell efficiently. There are several options here, transformation, transduction and conjugation. We are only concerned with conjugation and transduction in this report, which are extremely efficient in *Escherichia coli*. Conjugative systems occur in either a liquid environment or on a solid surface. Each system has its preference depending on its particular features as noted by Frost [2]. Second, the transposon must have a selectable marker that functions in the recipient. This may necessitate changing the promoter of the marker or even the marker itself depending on the recipient. Third, the transposon must reside on a suicide vector so that the carrier is not retained in the recipient cell. Finally, the delivery system must not adversely affect the recipient cell's physiology in any meaningful way. Depending on the ultimate goal, the delivery system should be matched with the species and strain phenotypes.

The system of this report begins with the plasmid pUT, an R6K-based replication plasmid derived from pGP704 [3]. The importance of this plasmid resides in the fact that the R6K origin can support plasmid replication only in a host strain that contains the R6K specific π protein of the *pir* gene required for replication. For any recipient strain not containing the *pir* gene, the plasmid will function as a suicide vector. A more thorough description of other suicide vector schemes is described by Herrero et al. [4]. An important feature of this plasmid is the RP4 *mob* (*oriT*) sequence that allows for high frequency conjugal transfer from the proper host strain. Another feature are the elements of a transposition system, the Tn*5* 19-bp terminal inverted repeats and a $IS50_R$ *tnp* gene modified to remove a *Not*I site oriented divergently from the I end. The pUTKm2 construct used in this exercise has a kanamycin resistance marker and additional cloning sites the most notable is the *Not*I site that is unique to the pUT vector backbone [1]. The donor host strain (BW19795) contains RP4-conjugative functions inserted into the *uidA* gene (*E. coli* β-glucuronidase) by recombination using M13 phage [5]. The donor host supports replication of a R6K replicon (*pir*$^+$) and is capable of conjugal transfer of plasmids containing the RP4 *oriT* sequences (*see* **Note 3**).

2 Materials

2.1 Strains and Bacteriophage

BW19795	RP4-2-tet::Mu-1*kan*::Tn*7*integrant/Δ*uidA*::*pir*$^+$ *hsdR*17 *srlC*300 *creB*510 *endA1 zbf*-5 *thi*	[5]
ISM1205	BW19795 pUTminiTn5Km2; Amp^R Kan^R	[1]
EDL933	*E. coli* serotype O157:H7:K- Nal^R	[6]

(continued)

LE392	K-12 F⁻ *hsdR514* (r_K^-, m_K^-) *glnV*(*supE44*) *tryT* (*supF58*) *lacY1* Δ(*lacIZY*)6 *galK2 galT22 metB1 trpR55* λ⁻	[7]
CSH50	Δ(*pro-lac*) *araBAD-0 rpsL*-(strR) *thi- fimE1*::IS*1*-	[8]
ISM200	LE392 Mu dI(Ap *lac*)*cts62*	F. C. Minion
Mu dI (Ap *lac*) *cts62*	Mu *c*ts62 d1(Ap^R *trp' B$^+$A'* Δ*W209-lacZYA*)	[9]

2.2 Media and Buffers

1. Luria–Bertani (LB) broth and agar per Liter: 10 g tryptone, 5 g yeast extract, 10 g NaCl; 15 g agar for plates. Mix and autoclave for 15 min at 121 °C. Cool agar to 55 °C in a water bath prior to pouring, 15 ml per 85 mm petri dish.
2. Mu phage buffer: LB + 2.5 mM Ca^{2+} and 2.5 mM Mg^{2+}.
3. Phage plates: LB + 2.5 mM Ca^{2+} and Mg^{2+}; the agar contains 12 g/l agar and plates are poured slightly deeper than normal LB plates.
4. Soft agar: LB with 6.5 g/l agar. Smaller volumes (100 ml) are usually prepared. It is indefinitely stable on the bench at room temperature.
5. Antibiotic concentrations are as follows: 100 μg/ml ampicillin, 50 μg/ml for kanamycin, and 20 μg/ml for nalidixic acid.
6. PBS (phosphate buffered saline) per liter: 8 g NaCl, 0.2 g KCl, 1.44 g Na_2HPO_4, and 0.24 g KH_2PO_4.
7. Sterile membranes, sterile filter or blotting paper.
8. Freezing media per liter: 10 g tryptone, 5 g yeast extract, 10 g NaCl, 6.27 g K_2HPO_4, 1.8 g KH_2PO_4, 0.5 g sodium citrate, 0.9 g $(NH_4)_2SO_4$, and 44 ml glycerol. Autoclave for 15 min at 121 °C, then add 0.4 ml of sterile 1 M $MgSO_4$ when cool.
9. Xgal (5-bromo-4-chloro-3-indoyl-β-D-galactopyranoside): prepare a stock solution of 10 mg/ml in dimethyl sulfoxide and store in aliquots at −20 °C for up to 6 months.

3 Methods

3.1 Conjugal Transposase Mutagenesis

1. Grow the conjugal donor strain (ISM1205) in LB broth plus 100 μg/ml ampicillin and 50 μg/ml kanamycin and the recipient (EDL933) in LB broth plus 20 μg/ml nalidixic acid overnight at 37 °C with shaking (*see* **Note 1**).
2. Wash both strains three times with PBS by centrifugation. This removes any antibiotics from the cultures.

3. Dilute 1 ml of cells with 5 ml of antibiotic-free LB and incubate at 37 °C with shaking until an OD_{600} of 0.7–1.0 is reached.
4. Transfer the donor strain to stationary conditions for 30 min to allow for regeneration of pili.
5. Prepare the mating surface by placing a sterile nylon membrane on a stack of sterile filter paper.
6. Combine 200 μl of each strain and plate the mating mixture on the membrane, allowing the liquid to be adsorbed by the filter paper by capillary action.
7. The sterile filter with the mating mixture is then transferred to an antibiotic-free LB plate cell side up.
8. Usually incubation occurs overnight at 37 °C but mating efficiency can be measured by taking mating mixtures at earlier time points (e.g., 2 h). We found it most convenient to just continue to the next morning.
9. Membranes are vortexed with 2 ml of LB broth and the mixture is incubated with shaking at 37 °C for 1 h.
10. 100 μl aliquots are spread-plated on LB plates containing 50 μg/ml kanamycin plus 20 μg/ml nalidixic acid for counter selection against the donor strain.
11. Each plate should have 300–400 colonies. These can be picked into 96-well plates with 150 μl of LB broth or freezing media with appropriate antibiotics per well and grown overnight at 37 °C for further analysis. Picking can be done using sterile toothpicks, but we normally use a Colony Picker (VP373 Colony Picker; V&P Scientific, Inc. San Diego, CA) to facilitate quicker transfer to 96-well plates.

    For analysis, each plate of 96 mutants is replica-plated using a 96-prong replicator in a fresh 96-well, flat bottom, nontreated polystyrene plate (Corning, Inc., New York, NY) containing 150 μl of LB broth per well. Assays should be developed to use these 96-well plates. Alternatively, a flat tooth 48 pin replicator (VP 407AH; V&P Scientific, Inc.) can be used to inoculate half a 96-well plate onto a standard 85 mm petri dish. It is best if this is done before freezing. Figure 1 shows a screen for biofilm-negative mutants in an *E. coli* EDL933 mini-Tn5Km2 library as described by Puttamreddy et al. [6].

    If colonies are picked to a freezing media, then the 96-well plates can be frozen following growth using an alcohol bath in a −70 °C freezer and stored indefinitely at −70 °C. If LB broth is used, glycerol will have to be added to a final concentration of 15%, mixed by shaking on a microtiter plate shaker prior to freezing.

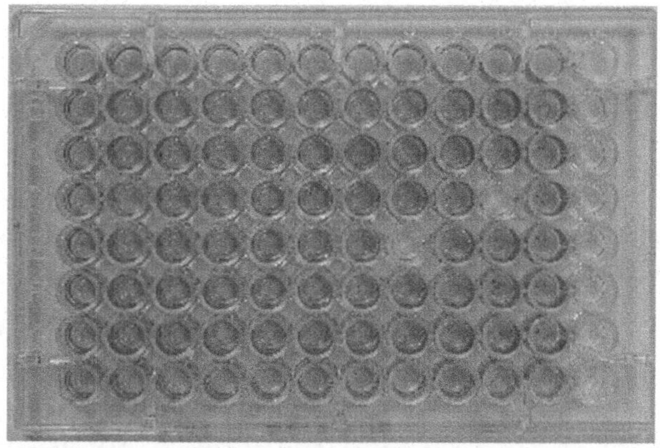

Fig. 1 Microtiter plate assay to screen for biofilm-negative mutants in *E. coli* 0157:H7 EDL933. The assay involved stationary growth of individual mutants in microtiter plates for 24 h at 37 °C. Plates were washed and stained with crystal violet. After further washing, dye was solubilized in 80% ethyl alcohol [6]. Wells in column 12 are controls. Wells D10 and E8 represent biofilm-negative mutants

3.2 Preparation of a Mu Lysate

1. The Mu lysogen is temperature sensitive and must be grown at 30 °C or lower at all times until induction. The lysogeny is struck out on a LB plate with 100 μg/ml ampicillin and grown overnight (or longer) at 30 °C until colonies are apparent and easily picked.
2. A single ampicillin resistant colony is picked into LB supplemented with 100 μg/ml ampicillin, 2.5 mM Ca^{2+} and 2.5 mM Mg^{2+} and grown overnight at 30 °C with shaking.
3. The overnight culture is diluted 1:100 in LB + Ca^{2+} and Mg^{2+} and grown at 30 °C with shaking for 3–4 h. The OD600 at this stage should be 1.0.
4. The flask is shifted to a water bath set at 42 °C and incubated for 35 min with shaking. Temperature is critical here to inactivate the Mu repressor *c*, which is temperature sensitive (*cts62*).
5. The flask is then incubated with shaking at 37 °C until the culture clears (~1.5 h).
6. The culture is transferred to a centrifuge tube, chloroform is added (0.25 ml) is added and the tube is vortexed to inactivate any remaining viable cells, and the tube is centrifuged at 5000 × *g* for 10 min. The supernatant contains the Mu d1 (Ap *lac*) phage.
7. The phage lysate is titered on phage plates with a typical *E. coli* K-12 strain (e.g., LE392). Dilute the phage in tenfold increments (10^{-3}–10^{-8}) in LB supplemented with Ca^{2+} and Mg^{2+}. This is usually done in 12 × 75 mm snap cap tubes. Add 190 μl of an overnight culture of *E. coli* K-12 with 10 μl of phage dilution, preadsorb for 25 min at 30 °C (or room temperature).

Add 3 ml of soft agar and plate on the phage plate. Grow at 42 °C overnight. This temperature is critical to prevent formation of a lysogen. Plaques are easily counted to determine the titer.

3.3 Generation of Gene Fusions with Mu d1(Ap lac)

1. Dilute the phage lysate to 10^{-3} and 10^{-4} in LB + Ca^{2+} and Mg^{2+}.
2. Grow a lac^{-} strain of *E. coli* (CSH50) overnight in LB + Ca^{2+} and Mg^{2+}.
3. Add 50 μl of diluted phage lysate to 250 μl of overnight growth and incubate the cell–phage mixture for 2 h at 30 °C.
4. Plate mixtures on LB + 100 μg/ml ampicillin and 50 μg/ml Xgal and grow at 30 °C. Sometimes this will take 2 days because of the lower growth temperature.
5. Blue and white colonies should be observed with various intensities of blue indicating the differences in transcription strength (*see* **Note 4**).

3.4 Rapid Sequencing of a Transposon Library

Once the library is constructed and is arrayed in 96-well plates, it will prove useful to design and implement a screening strategy. From that screen, a subset of mutants will be identified. Those mutants will need to be further characterized, generally by sequencing the insertion site. A couple of approaches come to mind. The approach covered here is one using PCR and Y linkers. Next Generation sequencing can also be used, particularly if the transposon contains a bar code as described by Wetmore et al. [10]. Using their TnSeq approach, one can locate every unique bar code to a chromosomal (or plasmid) location if the whole genome sequence is known. Then one only needs to know the bar code sequence that represents the mutant of interest. Conceivably, next generation sequencing could be used for the PCR-Y linker approach as well as long as an indexing primer is included in the approach.

1. Prepare Chromosomal DNA from 1 ml of mutant strain of interest (overnight grown). Quantify the DNA and use it for Y linker PCR immediately or store at −20 °C for further use.
2. For preparing the Y Linker, dilute Linker1 (TTTCTGCTC GAATTCAAGCTTCTAACGATGTACGGGGACACATG) & Linker 2 (TGTCCCCGTACATCGTTAGAACTACTC GTACCATCCACAT) [11] to 350 ng/μl each. Phosphorylate Linker2 using T4 polynucleotide kinase (PNK). Prepare a reaction of 9 μl linker2 [350 ng/μl], 2 μl of 10× PNK buffer, 2 μl 10 m M ATP, 1 μl T4 polynucleotide kinase, and 16 μl of nuclease-free water. Incubate at 37 °C for 10 min and then heat-inactivate the enzyme at 65 °C for 20 min). Then add 9 μl of linker 1. Heat to 95 °C for 2 min and cool slowly to room temperature. Analyze on gel. Quantify and store at −20 °C until use.

3. Prepare template by digesting chromosomal DNA with *Nla*III. Then ligate Y-Linker (CTGCTCGAATTCAAGCTTCT) to *Nla*III-digested chromosomal fragments. Prepare a mixture of 2.5 μl (500 ng) Y-Linker, 6.25 μl (50 ng) digested chromosomal DNA, 4 μl 5× T4 DNA ligase buffer, 1 μl T4 DNA ligase, and 7 μl nuclease-free water. Incubate over night at room temperature. Add 80 μl of nuclease-free water. Heat-inactivate at 65 °C for 10 min.
4. PCR Amplify the desired fragment using Tn5 (GGCCAGATCTGATCAAGAGA) and Y linker Primers [11]. For this prepare Master Mix 1 by mixing 21 μl nuclease-free water, 1 μl each of 10 μm Tn5 and Y linker primers, and 2 μl of Template DNA. Prepare Master Mix 2 separately by mixing 19.75 μl nuclease-free water, 5 μl of 10× PCR reaction buffer, 0.25 μl of Taq DNA polymerase (5 U/μl). Then mix both the master mixes and immediately start the cycle (94 °C 2 min, [94 °C 30 s, 58 °C 1 min, 72 °C 1 min] 30 cycles, 72 °C 5 min, and 20 °C hold.
5. Analyze on an agarose gel and sequence using Tn*5* and Y linker primers. Blast against the genome and identify the insertion location of transposon.

4 Notes

1. A variety of transposon derivatives are available for use in *E. coli* including Tn*5* and Tn*10* derivatives. For generating gene fusions, reporter genes can vary depending on the need and include β-galactosidase and alkaline phosphatase. Vectors are available to generate both transcriptional and translational fusions. Mini-transposon derivatives are available as well to enhance stability and provide for multiple rounds of mutagenesis because the transposase is not retained in the cell. Obviously, the use of transposons other than Tn*5* will necessitate the use of other primers for sequencing than that described above.
2. Mu bacteriophage is discussed because it replicates by transposition and in many aspects resembles transposons. Because of the availability of a temperature-sensitive repressor, Mu lysates are easily generated. The major drawback of this Mu derivative is its instability at higher temperatures because of the repressor mutation. Reports are that once inserted into the chromosome, MudI(Ap *lac*) is stable at 37 °C in a genetic background lacking a helper phage, but we have had our best results when maintaining lower temperatures.
3. We outline a way to deliver transposons to pathogenic *E. coli* through conjugation and transduction. However, other

possibilities exist using plasmids that are incompatible with the recipient strain. These would include those replicating using the R6K origin of replication. These plasmids only replicate in a *pir*-dependent strain. When *E. coli* is transformed with plasmids of this type containing a transposon, only those strains where the transposition has occurred to the chromosome become resistant. In conclusion, *E. coli* is a ready recipient of transposons and integrating phage (Mu). Most of the *Enterobacteriaceae* can be manipulated in a similar fashion.

4. If the particular strain under analysis is Lac^+, then the *lac* reporter in Mu d1(Ap *lac*) will not be useful for study. Other reporter constructs might be more useful and should be considered. Alternatively, one could generate a Lac^- strain by mutagenesis, and then the reporter constructs could be used to study gene regulation.

References

1. de Lorenzo V, Herrero M, Jakubzik U, Timmis KN (1990) Mini-Tn5 transposon derivatives for insertion mutagenesis, promoter probing, and chromosomal insertion of cloned DNA in gram-negative eubacteria. J Bacteriol 172 (11):6568–6572
2. Frost LS (2009) Conugation, Bacterial. In: Schaechter M (ed) Encyclopedia of microbiology, 3rd edn. Academic Press, Oxford, pp 517–531. https://doi.org/10.1016/B978-012373944-5.00007-9
3. Miller VL, Mekalanos JJ (1988) A novel suicide vector and its use in construction of insertion mutations: osmoregulation of outer membrane proteins and virulence determinants in *Vibrio cholerae* requires *toxR*. J Bacteriol 170:2575–2583
4. Herrero M, de Lorenzo V, Timmis KN (1990) Transposon vectors containing non-antibiotic resistance selection markers for cloning and stable chromosomal insertion of foreign genes in gram-negative bacteria. J Bacteriol 172 (11):6557–6567
5. Metcalf WW, Jiang W, Wanner BL (1994) Use of the *rep* technique for allele replacement to construct new *Escherichia coli* hosts for maintenance of R6Kγ origin plasmids at different copy numbers. Gene 138:1–7
6. Puttamreddy S, Cornick NA, Minion FC (2010) Genome-wide transposon mutagenesis reveals a role for pO157 genes in biofilm development in *Escherichia coli* O157:H7 EDL933. Infect Immun 78:2377–2384
7. Sambrook J, Fritsch EF, Maniatis T (1989) Molecular cloning: a laboratory manual, 2nd edn. Cold Spring Harbor Laboratory, Cold Spring Harbor, N. Y
8. Miller JH (1972) Experiments in molecular genetics. Cold Spring Harbor Laboratory, Cold Spring Harbor, N. Y
9. Casadaban MJ, Cohen SN (1979) Lactose genes fused to exogenous promotors in one step using a Mu-lac bacteriophage: In vivo probe for transcriptional control sequences. Proc Natl Acad Sci U S A 76(9):4530–4533
10. Wetmore KM, Proice MN, Waters RJ, Lamson JS, He J, Hoover CA, Blow MJ, Bristow J, Butland G, Arkin AP, Deutschbauer A (2016) Rapid quantification of mutant fitness in diverse bacteria by sequencing randomly bar-coded transposons. MBio 6(3):1–15
11. Kwon YM, Ricke SC (2000) Efficient amplification of multiple transposon-flanking sequences. J Microbiol Methods 41:195–199

Chapter 8

Methods for Transposon Mutagenesis in *Proteus mirabilis*

Philip N. Rather

Abstract

Several methods for transposon mutagenesis have been employed for use in *P. mirabilis*. The first method involves the use of mini-Tn*5* derivatives, which are delivered by conjugation of a suicide plasmid containing this transposon, followed by transposition into the chromosome. A second method is the use of preformed transposon/transposase complexes (transposomes), which are introduced into *P. mirabilis* cells by electroporation. Each of these methods will be discussed along with the advantages and disadvantages of each.

Key words *Proteus*, Transposon mutagenesis, Mini-Tn*5*, Transposome, Swarming, Virulence

1 Introduction

Proteus mirabilis is a human pathogen primarily associated with infections of the urinary tract [1–3]. This bacterium is noted for its robust swarming motility on agar surfaces [1, 4]. Genetic analysis in *Proteus mirabilis* has been greatly advanced by the use of transposon mutagenesis. In particular, transposon mutagenesis has been used to identify genes required for the process of swarming and for virulence [5–14]. The first example of transposon mutagenesis in *P. mirabilis* utilized a mini-Tn*5* derivative (mini-Tn*5Cm*) [15, 16]. This transposon contains the ends of Tn*5* together with an internal antibiotic resistance gene. This transposon is present on an R6K *pir*- suicide plasmid (pUT), where the Tn*5* transposase is encoded in trans. This plasmid can be delivered into *P. mirabilis* by conjugation using the appropriate *E. coli* donor strain (SM10, S17-1). Upon entering *P. mirabilis,* selection for the transposon resistance marker will identify cells with mini-Tn*5* insertions. A wide variety of mini-Tn*5* derivatives are available with different selectable markers and some have promoterless reporter genes such as *lacZ* for generating transcriptional or translational fusions [16]. A mini-Tn*5lacZ* derivative has been used successfully in *P. mirabilis* to generate random *lacZ* transcriptional fusions [6, 7]. In addition, signature-tagged mutagenesis has been

Steven C. Ricke et al. (eds.), *Microbial Transposon Mutagenesis: Protocols and Applications*, Methods in Molecular Biology, vol. 2016, https://doi.org/10.1007/978-1-4939-9570-7_8, © Springer Science+Business Media, LLC, part of Springer Nature 2019

reported using transposon derivatives [12, 13]. A significant drawback of the mini-Tn*5* system is that more than one insertion event can occur. In addition, transposition events are often associated with the integration of the pUT delivery plasmid at the site of transposon insertion. However, this method is highly efficient, and large numbers of insertions can easily be generated.

A second method utilizes commercially available complexes of Tn*5*-transposase complexes, designated transposomes. These complexes are electroporated into *P. mirabilis* cells where the transposon then inserts into the chromosome. The advantage of this system is that cells typically contain a single insertion, although the efficiency of generating insertions with this method is lower than with the mini-Tn5 system. In this chapter, both methods of transposon mutagenesis will be described.

2 Materials

1. Lysogeny broth growth media: 10 g tryptone, 5 g yeast extract, and 5 g NaCl per liter.
2. Antibiotics: Antibiotics are used at the following concentrations: *E. coli*; chloramphenicol (25 μg/ml), ampicillin (200 μg/ml), tetracycline (10 μg/ml), and kanamycin (20 μg/ml). For *P. mirabilis,* chloramphenicol is used at 35 μg/ml and kanamycin at 20 μg/ml.
3. LSW agar: Per liter: 10 g tryptone, 5 g yeast extract, 5 ml glycerol, 0.4 g sodium chloride, and 20 g agar.
4. Agar plates: Bacteria were grown on agar plates consisting of lysogeny broth containing 15 g agar per liter for *E. coli* and 30 g agar per liter to inhibit the swarming of *P. mirabilis.*
5. EZ-Tn5<KAN-2> transposon–transposase complex (Lucigen).
6. "TypeOne" Restriction inhibitor (Lucigen).
7. Bio-Rad cell electroporator.
8. 0.2 cm electroporation cuvettes.

3 Methods

3.1 Transposon Mutagenesis Using Mini-Tn5 Derivatives

1. Inoculate 2 ml LB broth containing 200 μg/ml ampicillin with *E. coli* SM10 λ*pir* or S17 λ*pir* containing a pUT plasmid with the desired mini-Tn5 transposon and shake overnight at 37 °C to stationary phase.
2. Inoculate 2 ml LB broth without antibiotics with *P. mirabilis* PM7002 (or any desired *P. mirabilis* strain) and shake overnight at 37 °C to stationary phase.

3. Pellet 1 ml of each culture in an Eppendorf microcentrifuge at 10,000 × *g* for 2 min, decant supernatant and wash with 1 ml of fresh LB broth. Repeat this wash two additional times.
4. Resuspend the final pellet in 500 μl LB.
5. In a separate Eppendorf tube, mix 50 μl of *E. coli* SM10 *λpir* containing the desired transposon on a pUT plasmid and 100 μl of the *P. mirabilis* strain to be mutagenized.
6. Spot 150 μl of the cell mixture on a well-dried 3% LB agar plate or LSW agar plate (*see* **Note 1**). Allow the "puddle" of cells to soak into the agar to create a concentrated mixture of cells. Do the same on separate plates with 50 μl of the *E. coli* SM10 donor and 100 μl of the *P. mirabilis* strain as controls. Incubate the plate for 6 h at 37 °C.
7. Add 4 ml LB broth to each plate and gently resuspend the cells using a glass spreader.
8. Transfer the resuspended mating mixture and the control cultures to a 15 ml conical tube.
9. Plate 100 μl and 200 μl of each culture on 3% LB agar plates or LSW agar plates containing 10 μg/ml tetracycline to counter-select the *E. coli* donor strain and either kanamycin (20 μg/ml) or chloramphenicol (35 μg/ml) depending on the mini-Tn*5* transposon used (*see* **Note 2**).
10. There should be at least a tenfold higher number of colonies on the plates containing the mating mixture versus each individual strain. Colonies of larger size often denote cells that contain more than one insertion (*see* **Note 3**).

3.2 Transposon Mutagenesis Using an EZ-Tn5<KAN-2> Transposome

1. Grow the desired *P. mirabilis* strain to be mutagenized in 25 ml of LB broth to an optical density of $A_{600} = 0.5$.
2. Pellet 2 tubes of 12.5 ml cells and resuspend in 1 ml of ice cold 10% glycerol.
3. Transfer cells to a 1.8 ml Eppendorf tube. Pellet cells in an Eppendorf microcentrifuge at 10,000 × *g* for 2 min, decant supernatant and wash with 1 ml of fresh LB broth. Repeat this wash two additional times.
4. Resuspend the final pellet in 200 μl ice cold 10% glycerol.
5. Add 1 μl of the EZ-Tn5<KAN-2> transposome complex and 1 μl of TypeOne restriction inhibitor to an Eppendorf tube. Add 70 μl of *P. mirabilis* cell mixture from **step 4**.
6. Immediately transfer to a 0.2 cm gap length cuvette and electroporate the mixture at 2.5 kV.
7. Immediately add 100 μl of prewarmed 37 °C LB media to the cells in the cuvette. Resuspend cells by gently pipetting up and down two times.

8. Transfer cells to 1 ml of prewarmed LB. Incubate stationary for 30 min and then shake at 37 °C 220 RPM for 1 h.
9. Plate aliquots on 3% LB agar plates or LSW agar plates containing 20 μg/ml kanamycin to select cells that have acquired EZ-Tn5<KAN-2> insertions.

4 Notes

1. LB plates must be dried extensively to allow cell mixtures to soak into the agar in a concentrated area. Dry plates inverted in a 37 °C incubator for 2 h before using.
2. Either LB with 3% agar or LSW agar can be used here. Both will prevent swarming and allow for the isolation of single colonies.
3. One disadvantage to using the mini-Tn*5* transposon is that cells occasionally get more than one insertion. When using chloramphenicol, larger colonies typically have more than one insertion. In addition, the pUT delivery plasmid will often insert into the chromosome along with the mini-Tn*5* element. This can be detected by colonies that are ampicillin resistant.

References

1. Armbruster CE, Mobley HL (2012) Merging mythology and morphology: the multifaceted lifestyle of *Proteus mirabilis*. Nat Rev Microbiol 10:743–754
2. Mobley HL, Belas R (1995) Swarming and pathogenicity of *Proteus mirabilis* in the urinary tract. Trends Microbiol 3:280–284
3. Armbruster CE, Mobley HLT, Pearson MM (2018) Pathogenesis of *Proteus mirabilis* infection. EcoSal Plus 8. https://doi.org/10.1128/ecosalplus.ESP-0009-2017
4. Morgenstein RM, Szostek B, Rather PN (2010) Regulation of gene expression during swarmer cell differentiation in *Proteus mirabilis*. FEMS Microbiol Rev 34:753–763
5. Belas R, Erskine D, Flaherty D (1991) *Proteus mirabilis* mutants defective in swarmer cell differentiation and multicellular behavior. J Bacteriol 173:6279–6288
6. Sturgill GM, Siddiqui S, Ding X, Pecora ND, Rather PN (2002) Isolation of lacZ fusions to *Proteus mirabilis* genes regulated by intercellular signaling: potential role for the sugar phosphotransferase (Pts) system in regulation. FEMS Microbiol Lett 217:43–50
7. Sturgill G, Rather PN (2004) Evidence that putrescine acts as an extracellular signal required for swarming in *Proteus mirabilis*. Mol Microbiol 51:437–446
8. Stevenson LG, Rather PN (2006) A novel gene involved in regulating the flagellar gene cascade in *Proteus mirabilis*. J Bacteriol 188:7830–7839
9. Clemmer KM, Rather PN (2007) Regulation of *flhDC* expression in *Proteus mirabilis*. Res Microbiol 158:295–302
10. Liaw SJ, Lai HC, Ho SW, Luh KT, Wang WB (2001) Characterisation of p-nitrophenylglycerol-resistant *Proteus mirabilis* super-swarming mutants. J Med Microbiol 50:1039–1048
11. Zhao H, Li X, Johnson DE, Mobley HL (1999) Identification of protease and rpoN-associated genes of uropathogenic *Proteus mirabilis* by negative selection in a mouse model of ascending urinary tract infection. Microbiology 145(Pt 1):185–195
12. Burall LS, Harro JM, Li X, Lockatell CV, Himpsl SD, Hebel JR, Johnson DE, Mobley HL (2004) *Proteus mirabilis* genes that contribute to pathogenesis of urinary tract infection: identification of 25 signature-tagged mutants attenuated at least 100-fold. Infect Immun 72:2922–2938

13. Himpsl SD, Lockatell CV, Hebel JR, Johnson DE, Mobley HL (2008) Identification of virulence determinants in uropathogenic *Proteus mirabilis* using signature-tagged mutagenesis. J Med Microbiol 57:1068–1078
14. Gibbs KA, Urbanowski ML, Greenberg EP (2008) Genetic determinants of self identity and social recognition in bacteria. Science 321:256–259
15. Belas R, Erskine D, Flaherty D (1991) Transposon mutagenesis in *Proteus mirabilis.* J Bacteriol 173:6289–6293
16. de Lorenzo V, Herrero M, Jakubzik U, Timmis KN (1990) Mini-Tn5 transposon derivatives for insertion mutagenesis, promoter probing, and chromosomal insertion of cloned DNA in Gram-negative eubacteria. J Bacteriol 172:6568–6572

Chapter 9

Mutagenesis of *Vibrio fischeri* and Other Marine Bacteria Using Hyperactive Mini-Tn*5* Derivatives

Julie L. Stoudenmire, Michael Black, Pat M. Fidopiastis, and Eric V. Stabb

Abstract

Mutagenizing bacterial genomes with selectable transposon insertions is an effective approach for identifying the genes underlying important phenotypes. Specific bacteria may require different tools and methods for effective transposon mutagenesis, and here we describe methods to mutagenize *Vibrio fischeri* using an engineered mini-Tn5 transposon with synthetic "mosaic" transposon ends. The transposon is delivered by conjugation on a plasmid that cannot replicate in *V. fischeri* and that encodes a hyperactive transposase outside the transposon itself. The chromosomal location of insertions can be readily identified by cloning and/or PCR-based methods described here. Although developed in *V. fischeri*, these tools and methods have proven effective in some other bacteria as well.

Key words *Vibrio*, *Photobacterium*, Transposon, Mutagenesis, Mini-Tn5

1 Introduction

Vibrio fischeri is a marine gammaproteobacterium best known for its symbiotic interactions and pheromone-controlled bioluminescence [1–7]. Initial studies employing transposon mutagenesis in *V. fischeri* showed the power of this approach but also had significant drawbacks, including relatively low transposon-insertion frequency, temperature sensitivity of the delivery vector, the occurrence of multiple insertions in a single strain, and the need to use a pleiotropic rifampicin-resistant *V. fischeri* mutant, rather than a true wild-type strain, to counterselect against *Escherichia coli* used as conjugative donors for the transposons [8, 9]. These limitations were largely overcome by new tools and methods for conjugation in *V. fischeri* [10] and by the development of a mini-Tn5-based transposon by Lyell et al. [11].

This mini-Tn5 system [11] took advantage of optimized 19-bp "mosaic" transposon ends [12, 13] along with a hyperactive transposase [14] to increase transposition frequency. Despite the increased transposition activity, mutants with multiple insertions

Steven C. Ricke et al. (eds.), *Microbial Transposon Mutagenesis: Protocols and Applications*, Methods in Molecular Biology, vol. 2016, https://doi.org/10.1007/978-1-4939-9570-7_9,

were not observed, keeping mutant analysis relatively straightforward [11]. Tn*5*-derived transposons generally lack target specificity, and while the location of insertion into DNA is not completely random, the small variance from randomness has few (if any) practical implications [15, 16]. Consistent with this track record for Tn*5* derivatives, the mini-Tn*5* system described here appears to deviate from truly random insertion-site selectivity, but this deviation from randomness is minor and does not affect transposon-mutagenesis strategies (unpublished data and [11, 17]).

The simplest transposon delivery vector in this lineage is pEVS170 (Fig. 1 and GenBank MH370733). The transposon in pEVS170 contains the *ermR* gene, which encodes a 23S rRNA methyltransferase and confers erythromycin resistance. Since its introduction to *V. fischeri* genetics by Visick and Ruby [18], the *ermR* erythromycin-resistance cassette has proven to be a superior selectable marker in this bacterium, due to the unambiguous resistance conferred by the *ermR* marker and the lack of background resistance or spontaneously erythromycin-resistant mutants. The transposon in pEVS170 also contains the R6K gamma origin of replication [19], which, as described below, allows the transposon and surrounding sequence to be recovered as a self-replicating plasmid in *E. coli* strains carrying the *pir* gene [19]. Another useful feature is an M13F priming site near one end of the transposon, which facilitates sequencing across transposon-insertion junctions. Since the initial description of pEVS170 [11], sequencing has revealed a three-base pair insertion relative to the published size of the transposon and plasmid (*see* corrected size in Fig. 1). Unique restriction sites within the transposon on pEVS170 have facilitated further engineering, as described below.

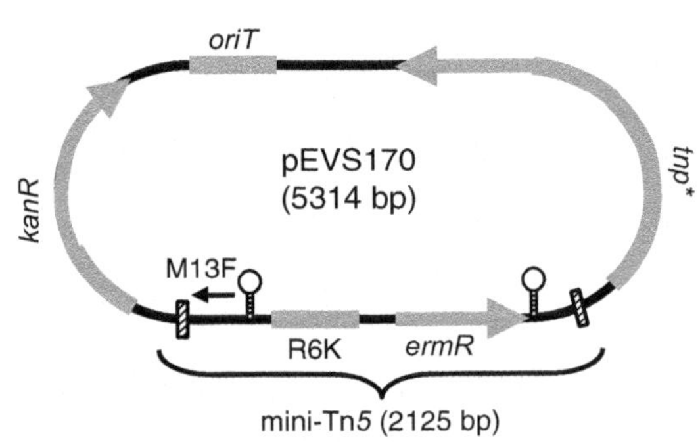

Fig. 1 Map of pEVS170. The plasmid includes a *tnp∗* gene, which encodes a hyperactive Tn*5* transposase, an RP4 origin of transfer (*oriT*), and a kanamycin resistance cassette (*kanR*). The hatched boxes represent the mosaic end sequences that define the mini-Tn*5* transposon. Included within the mini-Tn*5* transposon is the erythromycin resistance gene (*ermR*), Pi (*pir*)-dependent origin of replication (R6K), binding site for the M13 forward sequencing primer (M13F), and transcriptional terminators represented by stem loops

The pEVS170 vector sequence outside the transposon includes a mutant transposase gene [14], the RP4 origin of conjugative transfer (*oriT*), and a kanamycin resistance gene (*kanR*). As noted above, the mutant transposase gene enhances transposition. Specifically, an E54K change yields a more active transposase with improved binding to the transposon ends [14], and an M56A change eliminates an alternative translational start that can otherwise produce a truncated transposase protein with inhibitory activity [20]. The *oriT* supports transfer of the plasmid into target recipient strains with the assistance of RP4-based *tra* and *trb* conjugation machinery [21], whose components are supplied *in trans*. These conjugative functions have been provided by helper plasmid pEVS104 in triparental matings [10], although there are many other RP4-based conjugative donors available. The addition of *kanR* to the vector sequences outside the transposon proved important, because approximately 10% of *V. fischeri* recipients bearing the transposon were also kanamycin resistant [11], presumably due to unwanted transposition events that integrate the entire vector. The *kanR* marker provides an easy screen to identify and discard such mutants.

Variants similar to pEVS170 have been generated with other useful properties built into the transposons that they deliver (Fig. 2). For example, pEVS168 includes a promoterless *cat-gfp* two-gene transcriptional reporter encoding chloramphenicol resistance and green fluorescent protein [11]. More recently, two previously unpublished plasmids, pJLS108 and pHD1, were constructed with transposons that can generate translational fusions adding streptavidin-binding SA-1 tags to disrupted proteins when inserted in frame [22]. In a clever twist on insertional mutagenesis, Ondrey and Visick created pJMO10 with a transposon containing an outward-facing promoter ($P_{A1/34}$) [23] that is inducible with isopropyl β-D-1-thiogalactopyranoside (IPTG) provided *lacI* is also engineered into the recipient *V. fischeri* strain [24]. Thus, this construct can be used to identify genes of interest based on IPTG-dependent regulation of a gene adjacent to the site of insertion. Finally, pMJM10 [25] was engineered for use in transposon site hybridization (TraSH) screening [26], and it has promoters facing outward from the transposon ends that are specific for bacteriophage T7 RNA polymerase, as well as *Tsp*509I and *Mse*I restriction sites just outside the transposon, to facilitate the TraSH method. In the future, alternative selectable markers in the transposon could be useful, and *Eco*RV restriction sites flanking *ermR* would facilitate its exchange for other antibiotic-resistance genes (*see* pEVS170 sequence, GenBank MH370733).

Plasmid pEVS170 and its relatives have proven valuable in numerous studies of *V. fischeri* [17, 25, 27–36], but the results have also revealed points of caution worth noting. In one instance, transposon insertions failed to elicit the same phenotype when

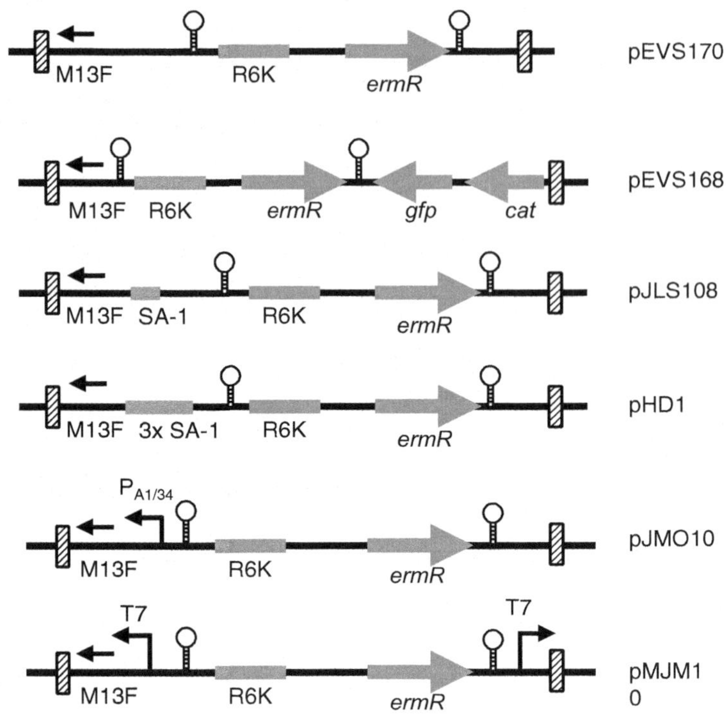

Fig. 2 Additional mini-Tn*5* transposon variants on plasmids similar to pEVS170. Graphical representations are similar to those in Fig. 1. Plasmid pEVS168 [11] carries a mini-Tn*5* with promoterless chloramphenicol resistance (*cat*) and green fluorescent protein (*gfp*) genes to form a transcriptional reporter upon insertion. Plasmids pJLS108 and pHD1 have transposons encoding SA-1 and 3× SA-1 tags, respectively. These transposons will generate translational fusions when inserted in the proper reading frame. Plasmid pJMO10 [24] has an outward facing $P_{A1/34}$ promoter that, in conjunction with a *lacI^q^* allele, allows for IPTG-inducible transcription across the transposon junction into flanking genes. The transposon in pMJM10 was engineered for use in TraSH screening [26] and has T7 promoters facing outward from both transposon ends to generate RNA in vitro with added T7 RNA polymerase. pMJM10 also has restriction sites engineered into the vector just outside the transposon ends [25], whereas the others are expected to be identical to pEVS170 outside the transposon borders

moved into a fresh genetic background. This unexpected result led to the discovery that spontaneous mutations in *luxO* give rise to strains that progressively dominate cultures of *V. fischeri* left in stationary phase too long [37]. In another study we discovered spontaneous small-deletion mutations in *celI* distant from the points of transposon insertion [17]. Such spontaneous *celI* mutants are easily detected in a blue–white screen, yet they have never been observed other than during transposon mutagenesis, suggesting that the procedure of transposon mutagenesis itself may increase the rate of other mutations. These observations have underscored the importance of genetically complementing transposon-insertion mutants to ensure that the insertion is causal to the phenotype. Shuttle vectors derived from a *V. fischeri* plasmid have been effective for this purpose [38].

Although developed for *V. fischeri* strain ES114 [39], the tools described here should be useful in other strains and bacterial species, and evidence is emerging that this is indeed the case. For example, pEVS170 or pEVS168 has been used effectively in *Vibrio cholerae* [40], *Vibrio parahaemolyticus* [41], *Vibrio salmonicida*, *Photobacterium leiognathi* [42], and *Ruegeria pomeroyi* (Alecia Septer, personal communication). It is likely that pEVS170 and the vectors in Fig. 2 will be useful in a variety of other bacteria as well.

2 Materials

2.1 Media

1. All media should be autoclaved (or otherwise sterilized) prior to use.
2. Lysogeny broth (LB) is prepared by adding 10 g tryptone, 5 g yeast extract, and 10 g NaCl to 1 L deionized water (dH_2O) and adjusting the pH to 7.4 using NaOH.
3. Lysogeny broth salt (LBS) is prepared by adding 50 mL of 1 M Tris–HCl pH 7.4–7.5, 10 g tryptone, 5 g yeast extract, and 20 g NaCl to 950 mL dH_2O.
4. For solid media (e.g., LB agar plates), add 15 g/L agar.
5. A kanamycin (kan) stock is prepared by dissolving 100 mg kanamycin per mL of dH_2O (e.g., 1 g in 10 mL) and filter sterilizing the resulting solution by passing it through a 0.2-μm filter. Aliquots of kan are frozen prior to use. Kan is added to LB (for selection of *E. coli*) at a concentration of 40 μg/mL and to LBS (for selection or screening of *V. fischeri*) at a concentration of 100 μg/mL.
6. Erythromycin (erm) is prepared by dissolving it at a concentration of 50 mg per mL into 70% ethanol. Such Erm stocks are stored in the freezer and added to LBS at a concentration of 5 μg/mL for selection of *V. fischeri*, or to brain heart infusion (BHI) at a concentration of 150 μg/mL for selection of *E. coli* (*see* **Note 1**).
7. Antibiotics are stored at −20 °C and added to agar-containing media after the media has been autoclaved and cooled in a 55 °C water bath.

2.2 Bacterial Strains

1. All bacterial strains (Table 1) are kept frozen at −80 °C in 20% glycerol stocks (made by mixing two parts 50% glycerol with three parts turbid culture).

Table 1
Bacterial strains used in this study[a]

Strain	Role in this method	Description or genotype
V. fischeri ES114	Recipient for Tn mutagenesis	Wild-type isolate from *Euprymna scolopes* light organ
E. coli CC118λpir	Host for conjugative helper plasmid pEVS104	Δ(*ara-leu*) *araD* Δ*lac74 galE galK phoA20 thi-1 rpsE rpsB argE*(Am) *recA λpir*
E. coli DH5αλpir	Host for pEVS170 and used for cloning R6K plasmids in low copy	φ80d*lacZ*ΔM15 Δ(*lacZYA-argF*)U169 *deoR supE44 hsdR17recA1 endA1 gyrA96 thi-1 relA1 λpir*
E. coli BW23474	Used for cloning R6K plasmids in high copy	Δ*lac-169 robA1 creC510 hsdR514 uidA* (Δ *Mlu*I):: *pir-116 endA* (*BT333*) *recA1*

[a]References for these strains are provided in the main text

2. Strains are streaked from −80 °C freezer stocks onto solid medium as described below, and isolated colonies are used to start cultures for the procedure. *E. coli* streak plates are incubated at 37 °C and *V. fischeri* is incubated at 28 °C.
3. *E. coli* CC118λpir [43] carrying conjugative-helper plasmid pEVS104 is streaked onto in LB + kan.
4. *E. coli* DH5αλpir [44] carrying transposon-donor plasmid pEVS170 (*see* **Note 2**) is streaked onto LB + kan.
5. *V. fischeri* ES114 (recipient strain) (*see* **Note 3**) is streaked onto LBS.
6. *E. coli* BW23474 [45, 46] is streaked onto LB (*see* **Note 4**).

2.3 Tn Mutant Identification—Method 1

1. PureLink Genomic DNA prep kit (Invitrogen).
2. DNA clean and concentrate kit (Zymo Research).
3. Quick Ligation Kit (New England Biolabs (NEB).
4. ZR plasmid miniprep kit—Classic (Zymo Research).
5. Restriction enzyme *Hha*I (NEB) and its corresponding reaction buffer stored at −20 °C.
6. Agarose.
7. 1× Tris-Borate-EDTA (TBE) buffer: For 5× TBE, add 54 g Tris Base, 27.5 g boric acid, and 20 mL 0.5 M EDTA pH 8.0 to 600 mL dH_2O and stir to mix and dissolve. Bring volume up to 1 L with dH_2O. Dilute a stock to 1× prior to use (*see* **Note 5**).
8. 1 KB DNA ladder (NEB).
9. 1000× ethidium bromide (0.5 mg/mL) (*see* **Note 6**).
10. 6× gel-loading dye (NEB).

11. Transformation-competent $CaCl_2$-treated *E. coli* BW23474 cells: In a 500 mL flask, inoculate 100 mL of LB with 1 mL overnight BW23474 culture. Grow cells to an OD_{600} of 0.4–0.5. Chill the cells on ice for 10 min. Centrifuge the cells at 4000 × *g* for 8 min at 4 °C. Discard supernatant. Gently resuspend cells in 50 mL ice-cold 100 mM $CaCl_2$ and incubate on ice for 5 min. Centrifuge cells at 4000 × *g* for 8 min at 4 °C. Discard supernatant. Gently resuspend the cells in 4 mL ice-cold 100 mM $CaCl_2$ with 20% glycerol. Place 100-μL aliquots of cell suspension into 0.5 mL microfuge tubes and leave at 4 °C for 12 h before using or moving to −80 °C for later use.

2.4 Tn Mutant Identification—Method 2

1. ZR Fungal/Bacterial gDNA MiniPrep kit (Zymo Research) (*see* **Note 7**).
2. DNA clean and concentrate kit (Zymo Research).
3. T4 DNA ligase (NEB) and its corresponding reaction buffer stored at −20 °C.
4. Restriction enzyme *Sau*3AI (NEB) and its corresponding reaction buffer stored at −20 °C.
5. Taq 2× Master Mix (NEB) stored at −20 °C.
6. DNA oligonucleotides (Table 2): VIPCR-F, VIPCR-RBIO, and MoSeq-F.
7. Agarose.
8. 1× Tris–borate–EDTA (TBE) buffer (*see* above).
9. 1 KB DNA ladder (NEB).
10. 1000× ethidium bromide (0.5 mg/mL).
11. 6× gel-loading dye (NEB).

3 Methods

3.1 Conjugation and Transposon Mutagenesis

1. From isolated colonies, grow overnight broth cultures of *E. coli* helper and donor strains (CC118λpir pEVS104 and DH5αλpir pEVS170) in LB + kan at 37 °C and the *V. fischeri* recipient strain (ES114) in LBS at 28 °C.
2. Combine 100 μL of each culture in a 1.5-mL microfuge tube and pellet the cells by centrifugation.
3. Wash the cell pellet with 1 mL fresh antibiotic-free LBS, and repellet the cells.
4. Completely remove the supernatant and resuspend the pellet in 10 μL fresh antibiotic-free LBS.
5. Spot this 10 μL onto a fresh LBS plate (*see* **Note 8**).

Table 2
Oligonucleotides used in this method

Name	5′–3′ sequence
M13F	TGT AAA ACG ACG GCC AG
VIPCR-F	CCT AGA GCG GCC GCA GA
[a]VIPCR-RBio	Biotin-ACT GGC CGT CGT TTT ACA G
MoSeq-F	AGA TGT GTA TAA GAG AC

[a]Oligonucleotide VIPCR-RBio should be 5′ biotinylated if sequencing the product using the PyroMark pyrosequencer

6. Incubate plate for ~16 h at a 28 °C.
7. Scrape the mating mix off the plate and resuspend in 500 μL LBS (*see* **Note 9**).
8. Plate the mating mix on selective plates (LBS + erm) (*see* **Note 10**) spreading 50–100 μL per plate and diluting in LBS if necessary (*see* **Note 11**).
9. Incubate plates at 22 °C 1–2 days (*see* **Note 12**).
10. Individual mutant *V. fischeri* colonies should be streak purified on LBS + erm to isolate them from other strains on the plate. *V. fischeri* colonies have a yellowish appearance that distinguishes them from *E. coli*.

3.2 Identification of Transposon Mutant Insertion: Method 1 (Direct-Cloning Method)

1. An overview of this method is shown in Fig. 3.
2. Individual transposon mutants of interest should be grown with shaking overnight in LBS broth with erm at 28 °C.
3. Isolate genomic DNA following the PureLink Genomic DNA prep kit instructions, eluting in dH_2O.
4. Digest 1 μg of genomic DNA with *Hha*I in a reaction volume of 50 μL, following the NEB enzyme protocol at 37 °C overnight (*see* **Note 13**).
5. Clean the digestion product, using the Zymo Clean and Concentrator kit, eluting with 15 μL dH_2O.
6. Set up a ligation reaction using the NEB Quick Ligation kit. 2.5 μL 2× Quick Ligase buffer, 1 μL cleaned *Hha*I-digested DNA (prepared above), 1.0 μL dH_2O, and 0.5 μL Quick Ligase enzyme. Allow reaction to proceed for 10 min at room temperature.
7. Transform the entire ligation reaction into $CaCl_2$-treated transformation-competent BW23474: Add the 5-μL ligation reaction to an aliquot of competent cells. Place the cells on ice for 30 min. Heat shock the cells in a 42 °C water bath for 45 s and place back on ice for 5 min. Gently transfer the cells to a 1.5 mL

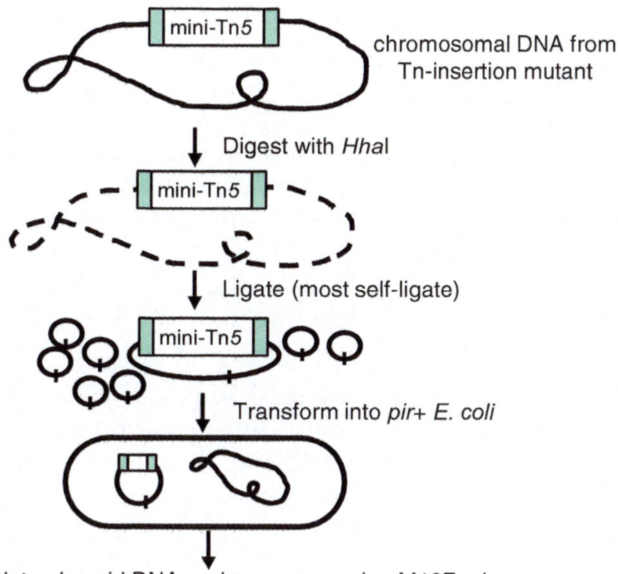

Fig. 3 Overview of Method 1 (direct cloning method) for identifying location of transposon insertion from chromosomal DNA. Chromosomal DNA from a mutant is digested with the restriction enzyme *Hha*I and self-ligated to generate a plasmid. The vertical line indicates a single *Hha*I site in the recircularized DNA. The ligation products are then transformed into *E. coli* containing the *pir* gene, allowing replication of the Tn-containing fragment by the Pi-dependent R6K origin of replication (*see* Fig. 1). Plasmid DNA is purified from *E. coli* clones that are erythromycin resistant and kanamycin sensitive, and these plasmids are sequenced using an M13F priming site just inside the transposon. Figure 4 shows representative sequencing results

microfuge tube and add 1 mL BHI. Incubate the tube with shaking at 37 °C for 1 h. Centrifuge the cells to pellet for 30 s and remove supernatant. Resuspend cells in 300 μL BHI and plate 100 μL each onto 3 prewarmed BHI + erm plates and incubate overnight at 37 °C.

8. Begin 3 mL cultures of transformants in BHI + erm and let grow overnight with shaking at 37 °C.
9. Miniprep the overnight cell culture using the ZR plasmid miniprep kit following manufacturer's instructions, eluting in 30 μL dH_2O.
10. Digest 5 μL of the miniprep sample with *Hha*I in a 20-μL reaction volume, following the NEB enzyme protocol for 1 h at 37 °C.
11. Run 10 μL of the digestion reaction on a 0.8% agarose gel: add 0.4 g agarose to 50 mL 1× TBE and microwave until the agar is dissolved (*see* **Note 14**). Allow the mixture to cool for approximately 10 min before adding 5 μL 10× ethidium

bromide to the mixture, swirl to mix and pour into electrophoresis cast. Allow the gel to set (approximately 20 min.) before adding to gel electrophoresis unit. Fill tank with 1× TBE, ensuring the entire gel is under the TBE level. Mix 10 μL digestion reaction with 2 μL 6× loading dye and load all 12 μL into a well on the gel. Load 2 μL uncut plasmid (mixed with 1 μL loading dye) next to the sample, and additionally load 2 μL 1Kb DNA ladder mixed with 1 μL loading dye. Run the gel, at an appropriate voltage for the electrophoresis unit, until the dye is three fourths of the way down the gel before imaging on a UV light box.

12. If a single band greater than 2.1 KB is identified in the digested sample, indicating only a single cut site, prep the sample for sequencing. If multiple bands are seen, then additional *Hha*I fragments have been cloned, and if there are many of them, the enlarged plasmid size may hinder downstream sequencing. If this problem appears, repeat **steps 5–10**, using 0.5 μL of the digested plasmid DNA and 1.5 μL dH_2O in **step 5**, until a single band is observed. In other words, digest again with *Hha*I and self-ligate to remove the additional *Hha*I fragments.
13. Sequence the miniprepped sample using the universal sequencing primer, M13F.
14. Search the sequence from **step 13** for the Tn mosaic end (5′ AGATGTGTATAAGAGACAG 3′). Typical sequencing results are shown in Fig. 4. For reference, the sequence between the M13F priming site and the transposon end is shown in Fig. 5. Search the sequence for the *Hha*I site (5′ GCGC 3′) where the regions upstream and downstream of the transposon were ligated together (Fig. 3). The sequence from the Tn–chromosome junction to the *Hha*I site can be mapped to the genome bioinformatically, for example using NCBI's BLAST search tool. Note that transposition results in a nine-base pair direct repeat of target DNA, so the first nine base pairs after the mosaic end will also be present just outside the other transposon end.

3.3 Identification of Transposon Mutant Insertion: Method 2 (Inverse PCR or iPCR Method)

1. The primers used in this method and their orientation relative to the transposon are shown in Fig. 5, and an overview of the method is illustrated in Fig. 6.
2. Individual transposon mutants of interest should be grown overnight in LBS erm at 28 °C.
3. Isolate genomic DNA following the ZR Fungal/Bacterial gDNA miniprep kit instructions, eluting in dH_2O.
4. Digest 1 μg of genomic DNA with *Sau*3AI in a reaction volume of 50 μL, following the manufacturer's protocol, for 1 h at 37 °C.

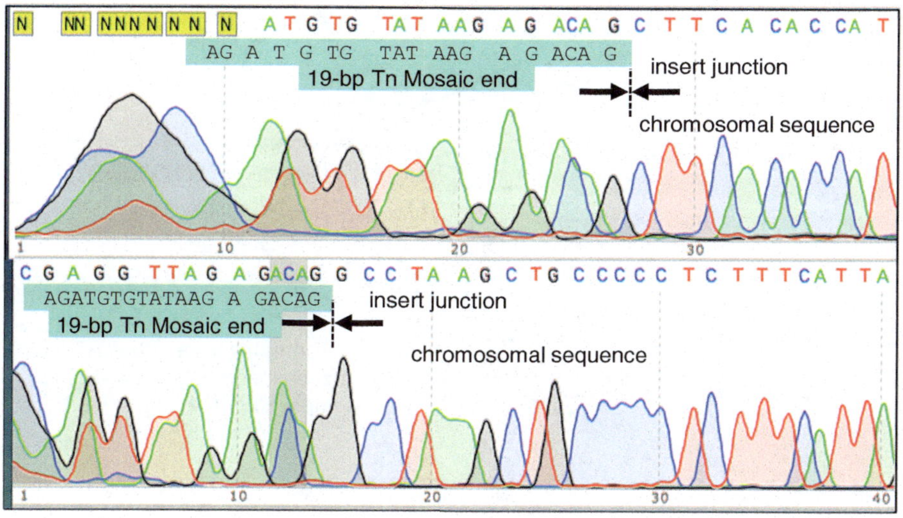

Fig. 4 Representative sequencing traces showing mini-Tn*5*–chromosome junctions. Following the cloning of transposon insertions and flanking DNA as self-replicating plasmids (Fig. 3), Sanger sequencing reveals the site of transposon insertion. Because the M13F sequencing primer is so close to the transposon end (*see* Fig. 5), the peaks can be crowded and automatic annotation may fail to identify Tn sequence. However, as shown in the two examples above, at least some of the 19-bp mosaic end of transposon usually be can identified in the sequencing traces, allowing for identification of the point of insertion. As with native Tn*5*, transposition results in a nine-base pair direct repeat of target DNA, and the first nine base pairs of chromosomal sequence shown above are also present at the other transposon–genome junction

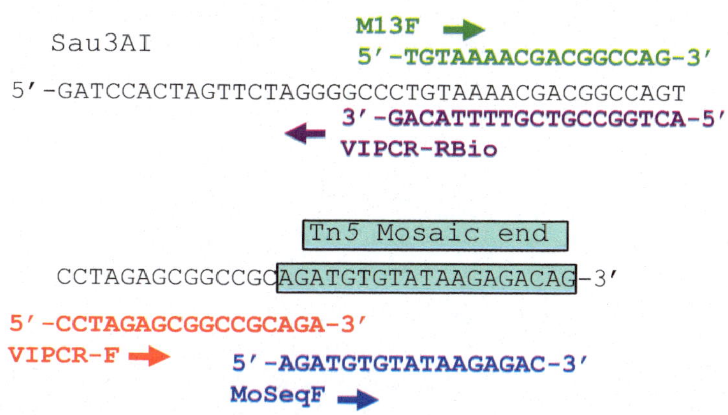

Fig. 5 Primer sequences and their alignment to the transposon in pEVS170. Sequence in black shows transposon sequence from one mosaic end (labeled and highlighted with cyan color) to the nearest *Sau*3AI site (5′-GATC-3′). The M13F sequencing primer is shown in green. The primers used in Method 2 for identifying the location of transposon insertions (*see* Fig. 6) are indicated and colored as VIPCR-RBIO (purple), VIPCR-F (red) and MoSeqF (blue)

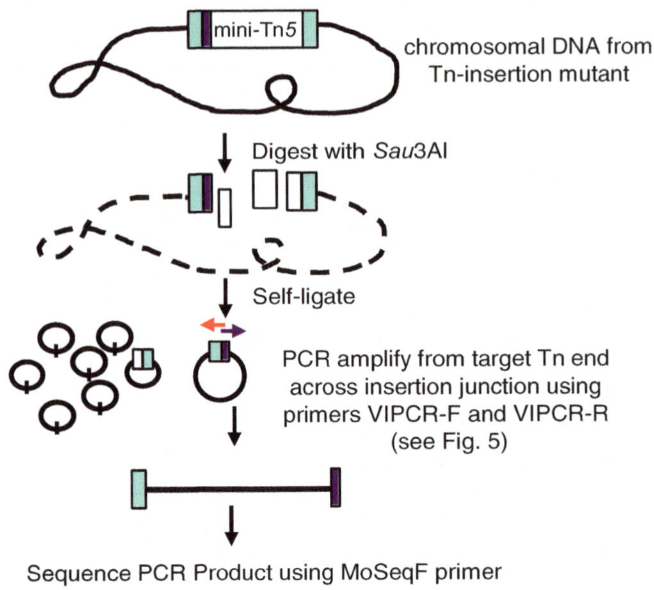

Fig. 6 Overview of Method 2 (inverse PCR or iPCR method) for identifying the location of transposon insertions from chromosomal DNA. Chromosomal DNA is digested with the restriction enzyme *Sau*3AI and self-ligated. The circular DNA is then PCR amplified with VIPCR-F and VIPCR-RBIO (depicted as red and purple arrows, respectively; *see* also Fig. 5). The resulting PCR product is sequenced using primer MoSeqF, which anneals to the mini-Tn*5* mosaic-end sequence (cyan color; *see* also Fig. 5)

5. Heat-inactivate the *Sau*3AI enzyme for 20 min at 70 °C.
6. Set up a ligation reaction using 25 μL of digested product in a total volume of 200 μL containing 5 Units of T4 DNA ligase in 1× T4 DNA ligase buffer.
7. Allow reaction to proceed for 2 h at room temperature.
8. Clean the ligation reaction using the Zymo Clean and Concentrator kit, eluting with 25 μL dH_2O (*see* **Note 15**).
9. Set up two PCR reactions with the purified ligation products, using 10 μL or 1 μL of the ligation mix as template, respectively. In each reaction, add 15 μL of 2× Taq master mix, primers (VIPCR-F and VIPCR-RBIO) to a final concentration of 0.2 μM each, and water to a total volume of 30 μL.
10. Load tubes into thermocycler with an initial denaturing step of 95 °C for 2 min and 45 cycles of 95 °C for 30 s, 60 °C for 30 s, and 72 °C for 30 s, followed by a final extension at 72 °C for 5 min. The reactions can then be held at 4 °C (*see* **Note 16**).
11. Run 5 μL of the PCR sample on a 2% agarose gel. Ideally there will be a single band from one or the other reaction that is less than 500 bp. Longer products may be more difficult to sequence using the PyroMark pyrosequencer.

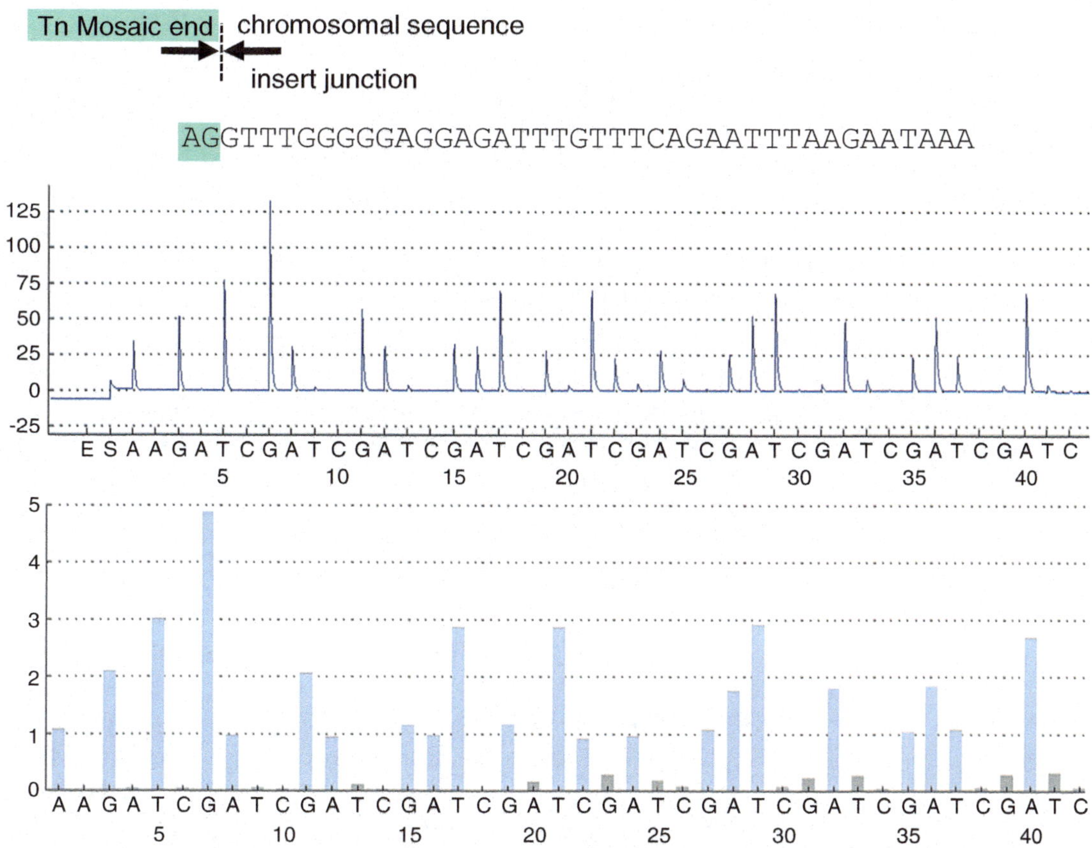

Fig. 7 Representative pyrosequencing result showing a mini-Tn*5*–chromosome junction. Following the iPCR method (Fig. 6), pyrosequencing with primer MoSeq-F (Fig. 5) usually captures the sequence immediately adjacent to the primer, beginning with the last two base pairs of the 19-bp mosaic end (5′-AG) followed by the Tn junction with genomic sequence. Panel A shows a pyrogram from sequencing an iPCR product with primer MoSeq-F, and panel B is a histogram showing the number of nucleotides incorporated with each dispensation. As with native Tn*5*, transposition results in a nine-base pair direct repeat of target DNA, and the first nine base pairs of chromosomal sequence shown above are also present at the other transposon–genome junction

12. Sequence the remaining PCR product using the primer MoSeq-F, following the PyroMark Q24 manufacturer's reagents and procedures, and a dispensation sequence of AA10(GATC) (*see* **Note 17**). A typical pyrosequencing result is shown in Fig. 7.
13. Genomic DNA sequence can be mapped to the genome with programs including NCBI's BLAST tool or similar resources.

4 Notes

1. BHI should be used (and not LB) when selecting erm resistance in *E. coli*. BHI is a complex medium sold in complete form by several suppliers. Anecdotal evidence from multiple

labs suggests that variations between batches of BHI may affect erm sensitivity of *E. coli*. Although such variation appears uncommon, it is prudent to ensure with any batch of BHI that *E. coli* carrying *ermR* is resistant to 150 μg/mL erm while other *E. coli* strains are sensitive.

2. Alternative transposon variants (Fig. 2) can be used, but the protocol for generating Tn mutants is the same. Donor strains bearing these plasmids (e.g., pEVS168) should likewise be streaked on LB with kan.

3. Other *V. fischeri* strains may be used as recipients in this protocol. As noted in the introduction, *V. cholerae*, *V. parahaemolyticus*, *V. salmonicida*, *P. leiognathi*, and *R. pomeroyi* have been used with this protocol as recipient strains, with some species-specific modifications for growth and selection. Molecular determination of transposition sites is the same regardless of recipient.

4. Other *pir*-containing strains can be used. BW23474 holds recovered plasmids at a higher copy number owing to the *pir-116* allele, and it generally results in higher plasmid yields in minipreps, facilitating sequencing.

5. Alternatively we have used TAE buffer for gel electrophoresis, replacing TBE with TAE throughout this protocol. 10× TAE can be prepared with (per liter final volume) 48.4 g Tris base (0.4 M), 11.4 mL acetic acid (0.2 M), 3.72 g EDTA (10 mM), and NaOH to adjust pH to 8.5. 10× TAE is then autoclaved, and upon dilution to 1× pH should be 8.3.

6. Ethidium bromide is a known mutagen and therefore should be treated with caution. Proper handling, including wearing gloves, and proper disposal are required. Alternative DNA stains suitable for agarose gel electrophoresis should work as well and are typically added after electrophoresis rather than added to the gel (*see* manufacturer's instructions for specific DNA stains). Similarly, ethidium bromide can be used to stain gels after electrophoresis, thereby not introducing this compound into the electrophoresis apparatus, if that is a concern.

7. Different genomic DNA prep kits are listed under the materials for Methods 1 and 2 to reflect how we have performed these procedures, but it is likely that either kit would work for either method.

8. Using fresh LBS plates increases the conjugation frequency (and therefore the number of transposon-insertion mutants) significantly relative to using older plates. Ideally, the plates should be fresh (poured within 24 h of use) but should not have standing puddles of water on the surface. If the plates are poured the same day, condensate on the agar surface can be dried by briefly exposing plates to a 37 °C incubator or oven,

provided plates are cooled back down prior to use. Alternatively, a sterile absorbent material can be used to remove excess moisture on the agar surface.

9. As an alternative method, the conjugation spot can be directly streak plated on selective media (LBS + erm), skipping this resuspension step and the next step of spread plating. Streak plating at this step is quicker but is likely to result in a lower total yield of transposon mutants.
10. An optional counter selection may be used to prevent *E. coli* growth. The addition of 2 μg/mL of potassium tellurite will prevent the growth of *E. coli* donors but allow *V. fischeri* recipients to grow. The *V. fischeri* colonies will have black centers, presumably owing to appearance of tellurium upon tellurite reduction. *V. fischeri* colonies may take slightly longer to appear in the presence of tellurite, but the *E. coli* background is effectively eliminated.
11. The procedure typically yields between 1000 and 10,000 transposon mutants per conjugation spot, so resuspending in 500 μL and plating 50–100 μL per plate should yield between 100 and 2000 CFU (transposon mutants) per plate. At the more successful end, additional dilution in LBS will be necessary to achieve isolated colonies.
12. Although *V. fischeri* grows well at 28 °C, incubating the selective plates following conjugation at this cooler temperature (22 °C) enriches more strongly for the *V. fischeri* recipient relative to the *E. coli* donor and helper strains.
13. *Hha*I has a four-base pair recognition sequence and cuts frequently in the *V. fischeri* genome but does not cut the transposon from pEVS170. Enzymes *Bst*UI and *Hin*PI are also four-base pair cutters and should also work for this purpose. *Bst*UI forms blunt ends, which may decrease the rate of self-ligation but should also decrease the appearance of additional cloned chromosome fragments.
14. Depending on the microwave, the time and power setting will vary. Boiling over should be avoided. Short pulses may be necessary to boil the buffer and dissolve the agarose without losing volume due to boil over.
15. As an alternative to this clean up, the ligation reaction can be ethanol precipitated by adding 20 μL of 3 M sodium acetate (pH 5) and 440 μL of 95% ice-cold ethanol for 10 min at −20 °C. The precipitated DNA is centrifuged for 10 min at 17,500 × *g* and the pellet is washed twice with 300 μL of 75% ethanol. The dried pellet is then resuspended in 25 μL of water.

16. These cycle times are used with Taq polymerase. Alternative PCR-compatible polymerases can be used but may require different time and temperature parameters.
17. Alternatively the PCR product can be cloned (and subsequently sequenced) using any of a number of commercially available kits (e.g., a Topo cloning kit from Invitrogen). However, note that the PCR product contains sequence for the M13F primer, which renders the M13F primer unsuitable for sequencing in situations where the vector also contains an M13F priming site.

Acknowledgments

The authors wish to thank Alecia Septer, Mark Mandel, Tim Miyashiro, and Karen Visick for helpful discussions, and Helen Dukes for technical assistance. This work was supported by NSF grants CAREER-MCB-0347317, IOS-0841480, IOS-1557964, IOS-0843317, and MCB-1716232.

References

1. McFall-Ngai MJ (2014) The importance of microbes in animal development: lessons from the squid-vibrio symbiosis. Annu Rev Microbiol 68:177–194. https://doi.org/10.1146/annurev-micro-091313-103654
2. Ruby EG, McFall-Ngai MJ (1992) A squid that glows in the night: development of an animal-bacterial mutualism. J Bacteriol 174:4865–4870
3. Stabb EV (2005) Shedding light on the bioluminescence "paradox". ASM News 71:223–229
4. Stabb EV (2006) The *Vibrio fischeri-Euprymna scolopes* light organ symbiosis. In: Thompson FL, Austin B, Swings J (eds) The biology of vibrios. ASM Press, Washington D.C., pp 204–218
5. Stabb EV, Schaefer A, Bose JL, Ruby EG (2008) Quorum signaling and symbiosis in the marine luminous bacterium *Vibrio fischeri*. In: Winans SC, Bassler BL (eds) Chemical communication among bacteria. ASM Press, Washington D.C., pp 233–250
6. Stabb EV, Visick KL (2013) *Vibrio fischeri:* a bioluminescent light-organ symbiont of the bobtail squid *Euprymna scolopes*. In: Rosenberg E, DeLong EF, Stackebrandt E, Lory S, Thompson F (eds) The prokaryotes, 4th edn. Springer-Verlag, Berlin Heidelberg, pp 497–532
7. Visick KL, Ruby EG (2006) *Vibrio fischeri* and its host: it takes two to tango. Curr Opin Microbiol 9(6):632–638. https://doi.org/10.1016/j.mib.2006.10.001. S1369-5274(06)00153-6 [pii]
8. Graf J, Dunlap PV, Ruby EG (1994) Effect of transposon-induced motility mutations on colonization of the host light organ by *Vibrio fischeri*. J Bacteriol 176:6986–6991
9. Graf J, Ruby EG (2000) Novel effects of a transposon insertion in the *Vibrio fischeri glnD* gene: defects in iron uptake and symbiotic persistence in addition to nitrogen utilization. Mol Microbiol 37:168–179
10. Stabb EV, Ruby EG (2002) RP4-based plasmids for conjugation between *Escherichia coli* and members of the Vibrionaceae. Methods Enzymol 358:413–426
11. Lyell NL, Dunn AK, Bose JL, Vescovi SL, Stabb EV (2008) Effective mutagenesis of *Vibrio fischeri* by using hyperactive mini-Tn*5* derivatives. Appl Environ Microbiol 74(22):7059–7063. https://doi.org/10.1128/AEM.01330-08. AEM.01330-08 [pii]
12. Reznikoff WS, Goryshin IY, Jendrisak JJ (2004) Tn*5* as a molecular genetics tool. In: Miller WJ, Capy P (eds) Mobile genetic elements: protocols and genomic applications, vol 260. Human Press Inc., Totowa, NJ

13. Zhou M, Bhasin A, Reznikoff WS (1998) Molecular genetic analysis of transposase-end DNA sequence recognition: cooperativity of three adjacent base-pairs in specific interaction with a mutant Tn*5* transposase. J Mol Biol 276 (5):913–925
14. Zhou M, Reznikoff WS (1997) Tn*5* transposase mutants that alter DNA binding specificity. J Mol Biol 271(3):362–373
15. Goryshin IY, Miller JA, Kil YV, Lanzov VA, Reznikoff WS (1998) Tn*5*/IS*50* target recognition. Proc Natl Acad Sci U S A 95 (18):10716–10721
16. Shevchenko Y, Bouffard GG, Butterfield YS, Blakesley RW, Hartley JL, Young AC, Marra MA, Jones SJ, Touchman JW, Green ED (2002) Systematic sequencing of cDNA clones using the transposon Tn*5*. Nucleic Acids Res 30(11):2469–2477
17. Adin DM, Visick KL, Stabb EV (2008) Identification of a cellobiose utilization gene cluster with cryptic beta-galactosidase activity in *Vibrio fischeri*. Appl Environ Microbiol 74 (13):4059–4069. https://doi.org/10.1128/AEM.00190-08. AEM.00190-08 [pii]
18. Visick KG, Ruby EG (1996) Construction and symbiotic competence of a *luxA*-deletion mutant of *Vibrio fischeri*. Gene 175:89–94
19. Kolter R, Inuzuka M, Helinski DR (1978) Trans-complementation-dependent replication of a low molecular weight origin fragment from plasmid R6K. Cell 15:1199–1208
20. Wiegand TW, Reznikoff WS (1992) Characterization of two hypertransposing Tn*5* mutants. J Bacteriol 174(4):1229–1239
21. Pansegrau W, Lanka E, Barth PT, Figurski DH, Guiney DG, Haas D, Helinski DR, Schwab H, Stanisich VA, Thomas CM (1994) Complete nucleotide sequence of Birmingham IncPa plasmids: compilation and comparative analysis. J Mol Biol 239:623–663
22. Bessette PH, Rice JJ, Daugherty PS (2004) Rapid isolation of high-affinity protein binding peptides using bacterial display. Protein Eng Des Sel 17(10):731–739. https://doi.org/10.1093/protein/gzh084
23. Bose JL, Rosenberg CS, Stabb EV (2008) Effects of *luxCDABEG* induction in *Vibrio fischeri*: enhancement of symbiotic colonization and conditional attenuation of growth in culture. Arch Microbiol 190:169–183
24. Ondrey JM, Visick KL (2014) Engineering *Vibrio fischeri* for inducible gene expression. Open Microbiol J 8:122–129. https://doi.org/10.2174/1874285801408010122
25. Brennan CA, Mandel MJ, Gyllborg MC, Thomasgard KA, Ruby EG (2013) Genetic determinants of swimming motility in the squid light-organ symbiont *Vibrio fischeri*. Microbiology 2(4):576–594. https://doi.org/10.1002/mbo3.96
26. Sassetti CM, Boyd DH, Rubin EJ (2001) Comprehensive identification of conditionally essential genes in mycobacteria. Proc Natl Acad Sci U S A 98(22):12712–12717. https://doi.org/10.1073/pnas.231275498
27. Bassis CM, Visick KL (2010) The cyclic-di-GMP phosphodiesterase BinA negatively regulates cellulose-containing biofilms in *Vibrio fischeri*. J Bacteriol 192(5):1269–1278. https://doi.org/10.1128/JB.01048-09
28. Lyell NL, Dunn AK, Bose JL, Stabb EV (2010) Bright mutants of *Vibrio fischeri* ES114 reveal conditions and regulators that control bioluminescence and expression of the *lux* operon. J Bacteriol 192(19):5103–5114. https://doi.org/10.1128/JB.00524-10
29. Lyell NL, Septer AN, Dunn AK, Duckett D, Stoudenmire JL, Stabb EV (2017) An expanded transposon mutant library reveals that *Vibrio fischeri* delta-aminolevulinate auxotrophs can colonize *Euprymna scolopes*. Appl Environ Microbiol 83(5). https://doi.org/10.1128/AEM.02470-16
30. Lyell NL, Stabb EV (2013) Symbiotic characterization of *Vibrio fischeri* ES114 mutants that display enhanced luminescence in culture. Appl Environ Microbiol 79(7):2480–2483. https://doi.org/10.1128/AEM.03111-12
31. Miyashiro T, Klein W, Oehlert D, Cao X, Schwartzman J, Ruby EG (2011) The *N*-acetyl-D-glucosamine repressor NagC of *Vibrio fischeri* facilitates colonization of *Euprymna scolopes*. Mol Microbiol 82(4):894–903. https://doi.org/10.1111/j.1365-2958.2011.07858.x
32. Miyashiro T, Oehlert D, Ray VA, Visick KL, Ruby EG (2014) The putative oligosaccharide translocase SypK connects biofilm formation with quorum signaling in *Vibrio fischeri*. Microbiology 3(6):836–848. https://doi.org/10.1002/mbo3.199
33. Post DM, Yu L, Krasity BC, Choudhury B, Mandel MJ, Brennan CA, Ruby EG, McFall-Ngai MJ, Gibson BW, Apicella MA (2012) O-antigen and core carbohydrate of *Vibrio fischeri* lipopolysaccharide: composition and analysis of their role in *Euprymna scolopes* light organ colonization. J Biol Chem 287 (11):8515–8530. https://doi.org/10.1074/jbc.M111.324012
34. Ray VA, Visick KL (2012) LuxU connects quorum sensing to biofilm formation in *Vibrio fischeri*. Mol Microbiol 86:954–970. https://doi.org/10.1111/mmi.12035

35. Studer SV, Mandel MJ, Ruby EG (2008) AinS quorum sensing regulates the *Vibrio fischeri* acetate switch. J Bacteriol 190:5915–5923. https://doi.org/10.1128/JB.00148-08. JB.00148-08 [pii]
36. Visick KL, Quirke KP, McEwen SM (2013) Arabinose induces pellicle formation by *Vibrio fischeri*. Appl Environ Microbiol 79 (6):2069–2080. https://doi.org/10.1128/AEM.03526-12
37. Kimbrough JH, Stabb EV (2015) Antisocial *luxO* mutants provide a stationary-phase survival advantage in *Vibrio fischeri* ES114. J Bacteriol 198(4):673–687. https://doi.org/10.1128/JB.00807-15
38. Dunn AK, Millikan DS, Adin DM, Bose JL, Stabb EV (2006) New *rfp*- and pES213-derived tools for analyzing symbiotic *Vibrio fischeri* reveal patterns of infection and *lux* expression in situ. Appl Environ Microbiol 72:802–810
39. Boettcher KJ, Ruby EG (1990) Depressed light emission by symbiotic *Vibrio fischeri* of the sepiolid squid *Euprymna scolopes*. J Bacteriol 172:3701–3706
40. Nuidate T, Tansila N, Saengkerdsub S, Kongreung J, Bakkiyaraj D, Vuddhakul V (2016) Role of indole production on virulence of *Vibrio cholerae* using *Galleria mellonella* larvae model. Indian J Microbiol 56(3):368–374. https://doi.org/10.1007/s12088-016-0592-6
41. Getz LJ, Thomas NA (2018) The transcriptional regulator HlyU positively regulates *exsA* expression leading to type III secretion system-1 activation in *Vibrio parahaemolyticus*. J Bacteriol 200(15):pii:e00653–pii:e00617. https://doi.org/10.1128/JB.00653-17
42. Dunn AK, Rader BA, Stabb EV, Mandel MJ (2015) Regulation of bioluminescence in *Photobacterium leiognathi* strain KNH6. J Bacteriol 197(23):3676–3685. https://doi.org/10.1128/JB.00524-15
43. Herrero M, De Lorenzo V, Timmis KN (1990) Transposon vectors containing non-antibiotic resistance selection markers for cloning and stable chromosomal insertion of foreign genes in gram-negative bacteria. J Bacteriol 172:6557–6567
44. Dunn AK, Martin MO, Stabb EV (2005) Characterization of pES213, a small mobilizable plasmid from *Vibrio fischeri*. Plasmid 54:114–134
45. Haldimann A, Prahalad MK, Fisher SL, Kim S-K, Walsh CT, Wanner BL (1996) Altered recognition mutants of the response regulator PhoB: a new genetic strategy for studying protein-protein interactions. Proc Natl Acad Sci U S A 93:14361–14366
46. Metcalf WW, Jiang W, Wanner BL (1994) Use of the rep technique for allele replacement to construct new *Escherichia coli* hosts for maintenance of R6K gamma origin plasmids at different copy numbers. Gene 138:1–7

Chapter 10

Transposon Mutagenesis of *Bacteroides fragilis*

Yaligara Veeranagouda, Fasahath Husain, and Hannah M. Wexler

Abstract

Bacteroides fragilis is Gram-negative obligatory anaerobe which usually resides in the gut of humans and animals. As an important member of the human gut microbiota it plays a vital role in digestion and absorption of nutrients as well as shaping of host immune system. *B. fragilis* is also infamous for causing serious infections. Treatment of *B. fragilis* infections caused emergence of multidrug-resistant strains. Molecular biology tools such as transposon mutagenesis help to decipher and understand commensal and pathogenic faces of *B. fragilis*. Using two mariner transposon vectors we describe the detailed methodology for the transposon mutagenesis of *B. fragilis*. We also describe two methods for the identification of transposon integration site (TIS) in transposon mutants. Transposon mutagenesis methods described in this chapter serve as a great tool for studying *B. fragilis*.

Key words Transposon mutagenesis, *Bacteroides fragilis*, Mariner transposon, Mutants

1 Introduction

Development of culture-independent identification of microorganisms has greatly enhanced our understanding of human gut microbiota. It is now an established fact that the gut microbiota plays a vital role in human biology [1, 2]. Recent developments in human microbiota studies clearly demonstrate that dysbiosis of gut microbiota is associated with diseases such as obesity, autoimmune disease, and neurological symptoms [1]. *B. fragilis* is an important member of human gut (gastrointestinal tract) microbiota. As a commensal, along with *B. thetaiotaomicron*, it contributes to the digestion of complex polysaccharides and development of host immune system [3, 4]. However, *B. fragilis* is also most frequently observed in sites of clinical infections such as tissue infections and bacteremia. As a pathogen it causes life-threatening infections [4]. In the past few decade treatment of *B. fragilis* infections with antibiotics leads to isolation of highly drug resistant strain from all over the world [5].

Steven C. Ricke et al. (eds.), *Microbial Transposon Mutagenesis: Protocols and Applications*, Methods in Molecular Biology, vol. 2016, https://doi.org/10.1007/978-1-4939-9570-7_10,

Finding new treatment strategies for *B. fragilis* infections requires a thorough understanding of genomic, transcriptomic, and physiology characteristics of *B. fragilis* and the mechanism of drug resistance [5]. In the past few years, high-throughput sequencing technologies such as next-generation sequencing (NGS) greatly contributed to our understanding of *B. fragilis* genomic diversity and pathogenic factors such as the diverse array of polysaccharide biosynthesis locus (PBL), multidrug efflux pumps, and drug resistance genes [3, 4, 6–9]. Tools such as transposon mutagenesis are also helpful for the construction of mutant libraries which can provide novel insights to gene function. In addition coupling of transposon mutagenesis tools with NGS for large-scale identification of mutants facilitates simultaneous investigation of thousands of genes [7, 10, 11]. Hence, transposon mutagenesis tools can also provide greater insights into *B. fragilis* physiology and pathogenicity.

The success of *B. fragilis* transposon mutagenesis projects depend on careful selection of a mutagenesis vector. A good transposon vector should exhibit the following characteristics: (1) easy mobilization to *B. fragilis*, (2) random insertion in to genome, (3) stable integration at a single genomic location without vector backbone integration, and (4) easy and efficient identification of transposon mutated gene. We previously reported the application of EZ::TN5 transposome for *B. fragilis* transposon mutagenesis [12]. Although EZ::TN5 transposome exhibited the aforementioned characteristics, high cost of EZ::TN5 transposome generation precluded this approach from routine application. This led to the search for cost-effective and efficient transposon mutagenesis tools for *B. fragilis*. At this end we explored the potential of pSAM-Bt, a mariner transposon mutagenesis vector which has been successfully used for the transposon mutagenesis of *B. thetaiotaomicron* and *P. gingivalis* [10, 11]. Using pSAM-Bt and its derivative pYV07 we successfully generated transposon mutants in *B. fragilis* 638R [13]. We recently demonstrated the use of transposon mutagenesis in the identification of essential genes and metronidazole resistance genes in *B. fragilis* 638R [7]. Here we describe the detailed method for transposon mutagenesis of *B. fragilis* and identification of transposon mutated genes by two different methods. The methodology described here can be helpful for researchers working on *B. fragilis*.

2 Materials

2.1 Transposon Mutagenesis and Mutant Isolation

1. *E. coli* S-17 λpir-pSAM_Bt (Prof. Andrew L. Goodman lab, [10].) or *E. coli* S-17 λpir-pYV07 (Prof. Hannah M Wexler lab; [13]).
2. *B. fragilis* 638R.

3. Luria–Bertani (LB) agar or LB broth (Sigma).
4. Brain–heart infusion (BHI) or BHI agar (Sigma) supplemented with 15 mg/l hemin (Sigma or Anaerobe Systems, Morgan Hill, CA, USA).
5. Anaerobic jars (Advanced instruments).
6. Anaerobic sachets (AnaeroPack CO_2) or Anoxomat Mark II or upper versions (Advanced Instruments) and gas cylinder containing 5% CO_2, 5% H_2, and 90% N_2.
7. Ampicillin (50 mg/l), kanamycin (40 mg/l), gentamicin (25 mg/l) and erythromycin (10 mg/l), rifampicin (10 mg/l).
8. Incubator.
9. Autoclave for sterilization of culture medium.
10. Pipette and nuclease-free tips.
11. Vortexer/multivial vortex shaker.
12. Spectrophotometer.
13. Glass test tubes (15 ml) and flasks (250 ml); sterile Falcon tubes (15 ml); petri dishes.
14. Sterile inoculating loops, needles, and spreaders (plastic or glass).
15. Glycerol.

2.2 Identification of Transposon Integration Site (TIS) in Mutant by Semirandom Primer Polymerase Chain Reaction (SRP-PCR)

1. PCR primers (*see* Table 1).
2. PCR tubes (0.2 ml).
3. OneTaq Hot Start 2× master mix (New England BioLabs, Inc.).
4. Thermocycler.

Table 1
Primers used in this study

Primer name	Sequence
SRP1	GGCCACGCGTCGACTAGTACNNNNNNNNNNGATAT
SAMSeq1	ACGTACTCATGGTTCATCCCGATA
SRP2	GGCCACGCGTCGACTAGTAC
SAMSeq2	GCGTATCGGTCTGTATATCAGCAA
SAMseq3	TCTATTCTCATCTTTCTGAGTCCAC
SRP3	GGCCACGCGTCGACTAGTACNNNNNNNNNNACGCC
SAMSeqR	GCCAGGCATCAAATTAAGCAG

5. PCR product cleaning kit (DNA Clean & Concentrator 25 kit, D4033, Zymo Research or Agencourt AMPure XP beads (A63880, Beckman Coulter, Inc.).

2.3 Identification of TIS in Mutants by Cloning

1. Genomic DNA isolation kit (ZR FUNGAL/BACTERIAL DNA MINIPREP, D6005 Zymo research).
2. Restriction digestion enzymes: BglII or EcoRI or PmeI or SacI or SphI (New England BioLabs, Inc.).
3. pUC19.
4. T4 DNA ligase (New England BioLabs Inc).
5. One Shot TOP10 Chemically Competent *E. coli* (Thermo-Fisher Scientific).
6. LB media.
7. 37 °C shaking and nonshaking incubator.
8. 42 °C water bath.
9. Ice bucket with ice.
10. LB-agar/kanamycin (40 mg/l) plates.
11. Plasmid purification kits.
12. Primers (*see* Table 1).

3 Methods

3.1 Transposon Mutagenesis of BF638R Using Transposon Vector

1. From frozen stocks, streak *E. coli* S-17 λ pir-pSAM-Bt (or pYV07) and *B. fragilis* 638R on LB-Agar + Ampicillin and BHI plates, respectively (*see* **Notes 1** and **2**).
2. Incubate LB + Ampicillin plates streaked with *E. coli* S-17 λ pir-pSAM-Bt (or pYV07) at 37 °C incubator. Transfer BHI plate containing *B. fragilis* 638R to anaerobic chambers. Create anaerobic atmosphere using anaerobic sachets or Anoxomat. Incubate the chamber at 37 °C incubator. Allow the cultures to grow for 16–24 h.
3. Next day evening, inoculate *E. coli* S-17 λ pir-pSAM-bt (or YV07) and *B. fragilis* 638R in 2.5 ml of sterilized LB + Ampicillin broth and BHI broth respectively. Transfer *B. fragilis* tubes into anaerobic chambers and create anaerobic atmosphere using anaerobic sachets or Anoxomat. Incubate the tubes at 37 °C for overnight.
4. Following morning, inoculate 50 μl *E. coli* S-17 λ pir-pSAMB-Bt (or pYV07) in 5 ml of sterilized LB media in 15 ml tubes plugged with cotton (2–3 tubes). For *B. fragilis* 638R, inoculate 5 ml of culture in 50 ml of sterilized BHI broth in 250 ml conical flasks plugged with cotton (two to three flasks).

Transfer *B. fragilis* 638R inoculated flasks to anaerobic chambers. Degas it using anaerobic sachets or Anoxomat *see* **Note 3**).

5. Incubates cultures at 37 °C for 3–4 h. At time interval, remove 1 ml of culture and measure optical density (OD) at 600 nm using spectrophotometer.
6. When culture OD reaches 0.1–0.3, place the culture tubes or flasks on ice (to avoid further growth) and mix 1 ml of *E. coli* S-17 λ pir-pSAM-Bt (or pYV07) with 10 ml of *B. fragilis* culture in 15 ml falcon tubes. Centrifuge culture broth at 5000 × *g* for 5 min (*see* **Note 4**).
7. Discard the culture supernatant and resuspend cell pellet in 100 μl of BHI broth.
8. Spread the resulting cell suspension on BHI plate and incubate the plates anaerobically at 37 °C for 3–5 h.
9. Remove plate from 37 °C and place it in anaerobic jars and create anaerobic atmosphere using anaerobic sachets or Anoxomat. Incubate the anaerobic jars at 37 °C for overnight (16–20 h).
10. Remove the plates from 37 °C, add 1 ml of sterile BHI + 10% glycerol broth to culture plate and scrape off the cells using a sterile spreader. This resulting cell suspension contains the mutant library.
11. Transfer the mutant library cell suspension to sterile 1.5 ml tubes.
12. Resulting mutant library can be used immediately for screening applications or store at −80 °C for long-term storage (6 months–1 year).
13. In order to determine transposon efficiency, remove the tube from −80 °C and thaw on ice and spread 20 μl of mutant library on BHI/Gentamycin (25 μg/ml)/erythromycin (10 μg/ml)/ plate. Transfer the plates to anaerobic jar and create anaerobic atmosphere by using anaerobic sachets or Anoxomat and incubate at 37 °C for 2–3 days (*see* **Note 5**).
14. Remaining mutant libraries can be stored at −80 °C for long-term storage (*see* **Note 6**).
15. For isolation of antibiotic-resistant or sensitive mutants, prepare BHI containing antibiotic of interest and plate 20–50 μl of mutant library.

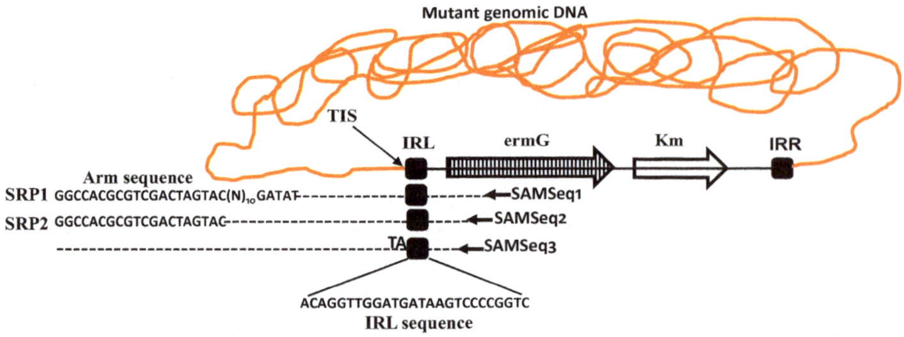

Fig. 1 Schematic representation of SPR-PCR. SPR-PCR is a type of nested PCR. In the first PCR cycle, SAMSeq1 primers bind to vector DNA, while the other one is a semirandom sequence (SRP1) which binds to random sequences on mutant DNA. SRP1 consists of an arm sequence, ten random bases (N) followed by five known bases (GATAT). The arm sequence serves as an anchor for second PCR while ten random bases facilitate random binding to mutant genomic DNA. A combination of six low annealing temperature (30 °C) cycles and thirty medium annealing temperature (45 °C) cycles facilitates amplification of transposon integration site (TIS). In second SRP-PCR, SAMSeq2 binds downstream of SAMSeq1, and SPR-2 binds to the arm region of SPR-1. Thus, the second PCR preferentially enriches TIS of mutant DNA. In addition, specificity of amplified PCR product can be increased by using sequencing primer (SAMSeq3) which can bind downstream of SAMSeq2. Bona fide sequence can be further confirmed by looking at IR sequence (IRL) and transposon integration site (TA). Sequence downstream of TA usually originates from TIS of mutant DNA

3.2 Identification of Transposon Integration Site (TIS) in Mutant by Semirandom Prime Polymerase Chain Reaction (SRP-PCR)

1. Using 20 μl pipette tips, transfer a portion of mutant colony into PCR tube containing 30 μl of water.
2. Place the tubes in thermocycler and run at 94 °C for 5 min.
3. Then cool the tube on ice, centrifuge at 13,000 × *g* for 1 min.
4. To perform SRP-PCR, transfer 23 μl of supernatant into a new PCR tube (Fig. 1). Add 1 μl of 10 μM of SRP1, 1 μl of 10 μM SAMSeq1 primers, and 25 μl of OneTaq Hot Start 2× master mix.
5. Place the tubes in a thermocycler and perform first-round PCR using settings shown in Table 2.
6. Transfer 1 μl of First PCR mix into new 0.2 ml PCR tube. Add 1 μl of 10 μM of SRP2, 1 μl of 10 μM SAMSeq2 primer, 22 μl water, and 25 μl of OneTaq Hot Start 2× master mix.
7. Place the tubes in a thermocycler and perform second-round SPR-PCR using settings shown in Table 3.
8. Purify PCR product using DNA Clean & Concentrator kit (Zymo research).
9. Sequence purified PCR product using SAMseq3 sequencing primer.
10. Identify the mutated gene by performing Blastn or Blastx analysis on the sequence (Fig. 2) (*see* **Note 7**). If the chromatogram fails to provide the sequence, repeat the above procedure by replacing SRP2 with SRP3 primer.

Table 2
First SRP-PCR conditions

Stage	Temperature (°C)	Time
Hold	95	2 min
Cycle (6 cycles)	95 30 68	30 s 30 s 1.5 min[a]
Cycle (30 cycles)	95	30 s
	45	30 s
	68	2 min[a]
Final extension	68	5
Hold	4	–

[a]5 s increment per cycle

Table 3
Second SPR-PCR conditions

Stage	Temperature (°C)	Time
Hold	95	10 min
Cycle (35 cycles)	95 55 68	45 s 45 s 1.5 min[a]
Final extension	68	10 min
Hold	4	–

[a]5 s increment per cycle

3.3 Identification of Transposon Integration Site (TIS) in Mutant by Cloning

1. Using Genomic DNA isolation kit prepare the genomic DNA (gDNA) from *B. fragilis* mutants generated by pYV07 (*see* **Note 8**).
2. Set up restriction digestion reaction for vector (pUC19) and gDNA in two separate tubes as follows:

CutSmart Buffer-10×	4
gDNA or pUC19	2–4 μg
EcoRI-HF	2
Water	make up volume to 40 μl

3. Incubate the reaction at 37 °C for 1 h.

A

TIS

>Met12-pSAMSeq3

CNNNNNNNNNNNNNNNNACNNNNAAATTTCGNNCGGCAATTGCTCGAGGGCGCGCCAAGCA
GAAGACGGCATACGAA**GACCGGGGACTTATCATCCAACCTGT**TAATACCATCAATCATCA
GATCCTTCAGCGGACCTTCAGACATATTATCCCTGATCAGATTCCCCAGTGCCTCAACGA
GCCATTCGATACCCATCATCGGATATTCACCGATCACAAAGGTACCTTCGAACATCAGAT
ACATAAAAAGAAAGAAGATGGGATATCCCC

B

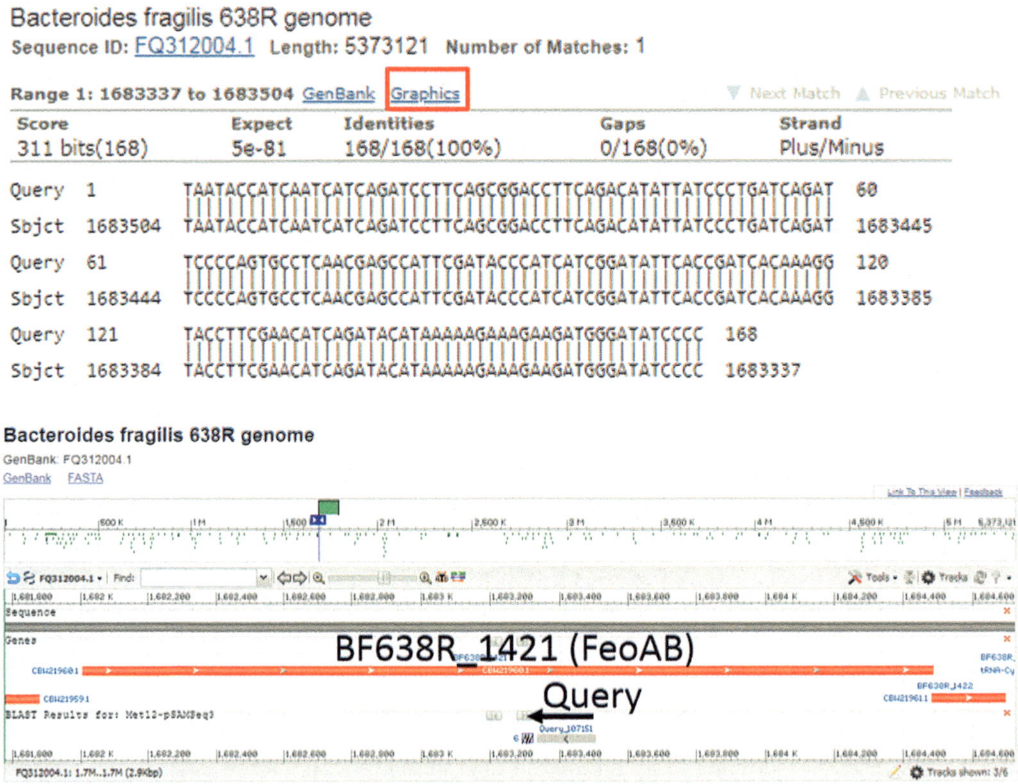

Fig. 2 Sequencing of TIS by SPR-PCR. This mutant is selected by plating transposon mutant library on BHI/Gentamycin (25 μg/ml)/erythromycin (10 μg/ml)/metronidazole (0.5 μg/ml)) plate. (**a**) An example of transposon integration site (TIS) obtained by sequencing of SRP-PCR product with SAM-Seq3 primer. Orange; vector sequence; Red = inverted repeat sequence in vector; Blue = TA site in mutant genome; Black; mutant DNA. (**b**) Blastn analysis of mutant DNA sequence. Mutant DNA mapped to *B. fragilis* 638R genome at 1683504 positions. Clicking on "Graphics" directs one to the corresponding genomic location of *B. fragilis* 638R genome. In the present case, transposon is integrated in the middle of BF68R_1421 gene (FeoAB). Disruption of BF638R_4121 leads to metronidazole resistance in *B. fragilis* 638R [8]

4. Remove gDNA containing tube and store at 4 °C. Add 3 μl of calf intestinal alkaline phosphatase (CIP) to pUC19 restriction digestion reaction mix and incubate at 37 °C for one more hour (*see* **Note 9**).
5. Purify both digestion mixtures using DNA Clean & Concentrator kit.

6. Set up ligation reaction as follows:

T4 DNA ligase buffer	2
Digested gDNA	0.3 μg
Digested pUC19	0.1 μg
T4 DNA ligase	1
Water	make up volume to 20 μl

7. Incubate at room temperature (22–25 °C) for 1 h. Then centrifuge the vial containing the ligation reaction briefly and place on ice.
8. Thaw one 50 μl vial of One Shot TOP10 Chemically Competent *E. coli* cells on ice. Pipet 20 μl of ligation reaction directly into the vial of competent cells and mix by tapping gently. Do not mix by pipetting up and down.
9. Incubate the vial(s) on ice for 30 min.
10. Incubate the tubes exactly 30 s in the 42 °C water bath. Do not mix or shake.
11. Remove tube from the 42 °C bath and place them on ice.
12. Add 250 μl of prewarmed S.O.C medium to each vial.
13. Place the tube in a microcentrifuge rack on its side and secure with tape to avoid loss of the tube. Shake the tube at 37 °C for exactly 1 h at 225 × *g* in a shaking incubator.
14. Configure tube at 1300 × *g* for 1 min, discard media without disturbing cell pellet. Resuspend the cells in 50 μl of S.O.C broth and spread it on LB/Kanamycin agar plates.
15. Invert the plate(s) and incubate at 37 °C overnight.
16. Streak 5–10 colonies on LB/Kanamycin agar plates and incubate at 37 °C overnight.
17. For quick screening, resuspend a small portion of cells in a tube containing 25 μl of water, 10 μl of phenol–chloroform solution, and 10 μl of 5× loading dye.
18. Vortex for 2 min and centrifuge at 13,000 × *g* for 5 min at room temperature. Prepare 1% agarose gel and load 10–20 μl of mix (from **step 17**) along with 1 kb ladder and run for 15–20.
19. Stain the gel with DNA staining agents and make sure that all clones exhibit a plasmid of similar size (*see* **Note 10**).
20. Then select the representative clone and prepare plasmid and perform sequencing using SAMSeq3 and SAMSeqR primer.
21. Identify mutated gene by performing Blastx analysis on sequence (*see* **Note 11**).

4 Notes

1. *B. fragilis* mutagenesis transposon vector pSAM-Bt or its derivative pYV07 are mariner transposon vectors. These vectors replicate only in *E. coli*. When these vectors are mobilized into *B. fragilis*, they work as suicide vectors and only the transposon within inverted repeats (containing erythromycin cassette) is integrated at TA site in *B. fragilis* genome. Both vectors carry ampicillin and erythromycin resistance cassette with promoters from *E. coli* and *B. fragilis* respectively. Ampicillin gene serves as the selection marker for *E. coli*, whereas erythromycin genes serve as the selection marker for *B. fragilis* transposon mutants. The pYV07 also has kanamycin resistance gene (with *E. coli* promoter) within transposon inverted repeats. An advantage of pYV07 is that it facilitates recovery of mutated gene by cloning.
2. *E. coli* S-17-1 λ pir strains contains the pir gene and conjugational transfer functions (RP4/RK6) in its genome. It facilitates biparental mating in lieu of triparental mating using helper strains.
3. Compared to *E. coli*, *B. fragilis* cultures grow slowly. So add more inoculum for *B. fragilis*. Maintain inoculum-to-media ratio at 1: 100 for *E. coli* and 1:20 for *B. fragilis*).
4. If project requires more mutant, one can mix *E. coli* S-17 λ pir-pSAMB-Bt (or pYV07) and *B. fragilis* 638R at 1:10 ratio in larger volumes. In subsequent steps, make sure to use one plate for 11 ml of culture).
5. *B. fragilis* 638R is resistant to gentamycin and sensitive to erythromycin. *E. coli* S-17 λ pir-pSAM-Bt (or pYV07) is resistant to ampicillin and sensitive to erythromycin and gentamycin. When mutant library is plated on BHI/Gentamycin (25 μg/ml)/erythromycin (10 μg/ml)/plate only transposon mutants (which gain erythromycin resistance by transposon integration) will survive. Depending on the OD of culture used at Subheading 3.1, **step 6**, 20 μl of mutant library yields 20–200 unique mutant colonies).
6. Mutant library retains its efficiency up to five freeze–thaw cycles.
7. The DNA sequence from second-round PCR product should contain IRL sequence (ACAGGTTGGATGATAAGTCCCCGGTCTT)). The presence of IRL is considered a characteristic of bona fide transposon mutants. The sequence next to IRL should start with TA (mariner transposon integration site). The remaining sequence can be identified by

performing Blastn or Blastx analysis against *B. fragilis* 638R genome.

8. Only mutants generated by pYV transposon vector can be identified by this method. In addition to erythromycin gene, transposon vector pYV07 also contains kanamycin cassette (which provides resistance to *E. coli*) within inverted repeat. Upon integration into host genome, erythromycin gene provides resistance to erythromycin and thus allows for selection of mutant from wild-type strains. Since kanamycin gene is active only in *E. coli*, cloning of digested genome in pUC19 vector facilitates selection on kanamycin.
9. CIP nonspecifically catalyzes the dephosphorylation of 5′ and 3′ ends of DNA. Thus, CIP treatment prevents self-ligation of linearized vector (pUC19) and increases cloning efficiency.
10. Usually all clones from a given mutant exhibit plasmids of similar size on gel. If the transposon integrates at more than one place in the genome, plasmids of two different sizes can be seen. In this case, sequence both clones.
11. The length of the obtained sequence depends on the occurrence of the restriction site on the genome. If the obtained length is not sufficient to retrieve the mutated gene, the process can be repeated by using different restriction enzymes. Restriction enzymes such as BglII, EcoRI, PmeI, SacI, and SphI do not cut integrated transposon sequence of pYV07; hence, these enzymes can be used for cutting mutant genomic DNA.

References

1. Sekirov I, Russell S, Antunes L (2010) Gut microbiota in health and disease. Physiol Rev 90:859–904. https://doi.org/10.1152/physrev.00045.2009
2. Round JL, Mazmanian SK (2009) The gut microbiota shapes intestinal immune responses during health and disease. Nat Rev Immunol 9:313–323. https://doi.org/10.1038/nri2515
3. Erturk-Hasdemir D, Kasper DL (2018) Finding a needle in a haystack: *Bacteroides fragilis* polysaccharide A as the archetypical symbiosis factor. Ann N Y Acad Sci 1417:116–129. https://doi.org/10.1111/nyas.13660
4. Wexler HM (2007) *Bacteroides*: The good, the bad, and the nitty-gritty. Clin Microbiol Rev 20:593–621. https://doi.org/10.1128/CMR.00008-07
5. Hecht DW (2004) Prevalence of antibiotic resistance in anaerobic bacteria: Worrisome developments. Clin Infect Dis 39:92–97. https://doi.org/10.1086/421558
6. Husain F, Veeranagouda Y, Hsi J et al (2013) Two multidrug-resistant clinical isolates of *bacteroides fragilis* carry a novel metronidazole resistance nim gene (nimJ). Antimicrob Agents Chemother 57:3767–3774. https://doi.org/10.1128/AAC.00386-13
7. Veeranagouda Y, Husain F, Tenorio EL, Wexler HM (2014) Identification of genes required for the survival of *B. fragilis* using massive parallel sequencing of a saturated transposon mutant library. BMC Genomics 15:429. https://doi.org/10.1186/1471-2164-15-429
8. Veeranagouda Y, Husain F, Boente R et al (2014) Deficiency of the ferrous iron transporter FeoAB is linked with metronidazole resistance in *Bacteroides fragilis*. J Antimicrob Chemother 69:2634–2643. https://doi.org/10.1093/jac/dku219
9. Rashidan M, Azimirad M, Alebouyeh M et al (2018) Detection of *B. fragilis* group and

diversity of bft enterotoxin and antibiotic resistance markers cepA, cfiA and nim among intestinal *Bacteroides fragilis* strains in patients with inflammatory bowel disease. Anaerobe 50:93–100. https://doi.org/10.1016/j.anaerobe.2018.02.005

10. Goodman AL, McNulty NP, Zhao Y et al (2009) Identifying genetic determinants needed to establish a human gut symbiont in its habitat. Cell Host Microbe 6:279–289
11. Klein BA, Tenorio EL, Lazinski DW et al (2012) Identification of essential genes of the periodontal pathogen *Porphyromonas gingivalis*. BMC Genomics 13:578. https://doi.org/10.1186/1471-2164-13-578
12. Veeranagouda Y, Husain F, Wexler HM (2012) Transposon mutagenesis of the anaerobic commensal, *Bacteroides fragilis*, using the EZ::TN5 transposome. FEMS Microbiol Lett 333:94–100
13. Veeranagouda Y, Husain F, Wexler HM (2013) Transposon mutagenesis of *Bacteroides fragilis* using a mariner transposon vector. Anaerobe 22:126–129. https://doi.org/10.1016/j.anaerobe.2013.04.012

Chapter 11

Transposon Mutagenesis in *Mycobacterium avium* Subspecies *Paratuberculosis*

John P. Bannantine, Denise K. Zinniel, and Raúl G. Barletta

Abstract

While transposon mutagenesis has been developed for *Mycobacterium avium* subspecies *paratuberculosis* (*Map*), relatively few laboratories have adopted this important genetic tool to examine gene function and essentiality. Here we describe the construction of a *Map* transposon library using the *Himar1* mariner transposon, but concepts can also be applied to the Tn*5367* transposon, which has also been used by our group. Delivery of the transposon is by a temperature-sensitive phagemid, ϕMycoMarT7, and plating transductants requires patience and specialized media due to length of incubation required to observe colonies. Several transposon mutants obtained from these libraries have been tested in vaccine and pathogenesis studies. By providing the following detailed protocol herein, we expect to demystify the procedure and encourage additional investigators to incorporate transposon mutagenesis in their studies on Johne's disease.

Key words Transposon mutagenesis, Mycobacterium, Johne's disease

1 Introduction

Mycobacterium avium subspecies *paratuberculosis* (*Map*) causes Johne's disease in cattle, sheep, goats, and other ruminants. It is among the slowest growing mycobacteria with a generation time of approximately 24 h and some strains, especially those isolated from sheep, are difficult to culture. These factors have combined to delay advancements in genetic studies of *Map* relative to other mycobacteria including *M. smegmatis* and *M. tuberculosis*, especially as it relates to vaccine strain construction and essential gene analysis.

The first efforts to genetically modify *Map* came with the discovery that mycobacteriophage TM4 could form plaques on *Map* lawns [1]. Soon after, a TM4 derivative, phAE94, was engineered to deliver the Tn*5367* transposon randomly into the *Map* genome [2]. Tn*5367* is derived from IS*1096* from *M. smegmatis* [3]. With this technology in hand, a library of over 5000 transposon mutants was initially constructed [2], and later, 13,500 Tn*5367*

Steven C. Ricke et al. (eds.), *Microbial Transposon Mutagenesis: Protocols and Applications*, Methods in Molecular Biology, vol. 2016, https://doi.org/10.1007/978-1-4939-9570-7_11,

mutants of *Map* were obtained [4]. Directed knockouts using insertion sequences have now also been achieved in *Map*. Examples of genes that have been targeted using this method include *sigH* [5, 6], *lipN* [5], *leuD* [7], *relA*, *lsr2*, and *pknG* [8], with *relA* [9] and *leuD* [10] mutants showing protection from *Map* challenge in goats.

More recently, our group compared the total genome insertions sites for the mycobacterial transposon Tn*5367* with the mariner transposon MycoMarT7 carrying the *Himar1* transposase [11]. The dinucleotide insertion recognition site (5′-TA-3′) for MycoMarT7 was significantly more prevalent than the degenerate tetramer (5′-NNPy[A/T]A[A/T]NN-3′) Tn*5367* recognition sites such that only 37 genes do not contain a MycoMarT7 site, compared to 710 genes missing Tn*5367* sites. Therefore, transposons containing the IS*1096* element have recognition site distribution biases that could affect the interpretation of gene essentiality on a whole genome basis. Nonetheless, this transposon has been useful for single gene studies and mutant library selection under defined conditions [4, 12–15].

Tn*5367* carries a kanamycin-resistance marker and can be delivered to *Map* by a thermosensitive phagemid phAE94 derived from mycobacteriophage TM4 [1, 16]. Whereas, MycoMarT7 is a *Himar1*-derived transposon that also carries a kanamycin-resistance marker engineered into the same TM4 phagemid, designated ϕMycoMarT7 [17, 18]. Here we describe the detailed protocols for obtaining a *Himar1* transposon library in *Map*.

2 Materials

For transposon mutagenesis, the only two strains of *Map* used thus far are the type strain (ATCC19698) and the bovine clinical isolate K-10. Other *Map* stains are expected to be amendable to the genetic manipulations described here, although ovine isolates will be more fastidious to grow in the laboratory [19].

1. Middlebrook 7H9 medium: Dissolve 4.7 g of Difco™ 271310 Middlebrook 7H9 powder (Becton, Dickinson and Company) in 900 ml distilled H_2O and 2 ml of glycerol. For *Map*, use HCl to adjust the pH to 5.9. If preparing culture media for petri dish applications, microbiological agar (15 g/l) is added. Autoclave at 121 °C for 20 min with slow exhaust. Cool to 50 °C (*see* **Note 1**) in order to add supplements (e.g., Tween™ 80, cycloheximide, albumin–dextrose–catalase (ADC; BBL™ Middlebrook 212352; Becton, Dickinson and Company), oleic acid–albumin–dextrose–catalase (OADC; BBL™ Middlebrook 212240; Becton, Dickinson and Company), mycobactin J (Allied Monitor, Inc.)) and any desired antibiotic(s) for

selection. Under sterile conditions, add 2.5 ml of 20% Tween™ 80, 10 ml of 1% cycloheximide and 100 ml of ADC (*M. smegmatis*). For *Map*, add 2.5 ml of 20% Tween™ 80, 10 ml of 1% cycloheximide, 1 mg of mycobactin J, and 100 ml of OADC instead of ADC. Tween™ 80 is not added if preparing agar (*see* **Note 2**).

2. Middlebrook 7H10 agar medium: Dissolve 19 g of Difco™ 262710 Middlebrook 7H10 powder (Becton, Dickinson and Company) in 900 ml distilled H_2O and 5 ml of glycerol. Autoclave at 121 °C for 20 min with slow exhaust. Cool to 50 °C and add supplements and/or antibiotic(s) as listed above (*see* Subheading 2, **item 1**).
3. Mycobacterial phage (MP) buffer: For *M. smegmatis*, the standard composition is 50 mM Tris–HCl, pH 7.5; 150 mM NaCl; 10 mM $MgSO_4{\cdot}7H_2O$; and 2 mM $CaCl_2$. For *Map*, the components are 50 mM Tris–HCl, pH 7.6; 150 mM NaCl; 10 mM $MgCl_2$; and 2 mM $CaCl_2$. These both can be stored up to 1 year at room temperature.
4. Phage soft agar: media is prepared as stated above (*see* Subheading 2, **item 1** or **2**) depending on the bacterial source except that 7 g/l of microbiological agar is added.
5. Adsorption-stop buffer: MP buffer containing 20 mM sodium citrate and 0.2% Tween™ 80. This solution is added to prevent further phage infections after completion of the adsorption time.
6. 50% glycerol-PBS freezing solution: 2 mM NaH_2PO_4, 4 mM Na_2HPO_4, 4 mM NaCl, and 50% glycerol (vol/vol).

3 Methods

3.1 Mycobacterial Culture

1. Inoculate a 2-l flask containing 500 ml of Middlebrook 7H9 at pH 5.9 supplemented with 0.2% glycerol, OADC, 0.05% Tween™ 80, 1 mg mycobactin J and 0.01% cycloheximide (*see* Subheading 2, **item 1**). Use a single, well isolated colony of the *Map* strain of choice for the inoculum.
2. Incubate culture stagnant for 4 weeks checking optical density at 540 nanometers (O.D. $_{540nm}$) after the third and fourth week.
3. Harvest 50-ml aliquots of cultures in mid-log phase (O.D.$_{540\ nm}$ = 0.4 to 0.75) by centrifugation at 4420 × *g* for 25 min. This optical density equates to approximately 2.0×10^8 cfu/ml.
4. If preparing a fresh stock of the phagemid, start the phagemid preparation protocol (Subheading 3.2) during the fourth week of growing the *Map* culture.

3.2 Phagemid Preparation

The phagemid ϕMycoMarT7 must be propagated and titered using freshly grown *M. smegmatis* cells at 30 °C.

1. Grow approximately 10 ml of *M. smegmatis* mc^{2}155 shaking overnight at 37 °C in Middlebrook 7H9 media supplemented with 0.2% glycerol, ADC, 0.05% Tween™ 80 and 0.01% cycloheximide (*see* Subheading 2, **item 1**).
2. From a known ϕMycoMarT7 titer number, make appropriate dilutions in MP buffer. For example, phagemid at a titer of 10^{10} plaque-forming units (PFU)/ml will usually have a "lace" pattern (plaques are touching) at 10^{-6} that is used for propagation and countable plaques are determined at the 10^{-8} dilution. The final dilution should be made several times depending on how much phage needs to be propagated (*see* **Note 3**).
3. Mix the *M. smegmatis* mc^{2}155 culture with the phagemid dilution in a 2:1 volume ratio (e.g., 2 ml of culture with 1 ml diluted phagemid).
4. Incubate standing at 30 °C for 30 min to allow for adsorption.
5. Add 300 μl of the phagemid–cell mixture to 3.5 ml of phage soft agar (*see* Subheading 2, **item 4**) and gently mix by rolling the tube with your palms. It is important to keep the phage soft agar at 56 °C until ready to use in a glass tube in a heat block.
6. Pour each tube of phage soft agar containing phagemid–cell mixture onto separate Middlebrook 7H9 agar plates (*see* **Note 4**). Quickly rotate the plate in a circular motion to evenly spread the agar over the entire surface. Once solidified, invert the plates to prevent condensation.
7. Incubate at 30 °C overnight or until plaques are visible (2–3 days).
8. Once plaques are produced and clearly visible, flood each plate with 5 ml of MP buffer. These plates are then incubated at 4 °C for 2–4 h followed by incubation at 37 °C for 1–2 h (*see* **Note 5**).
9. Collect the liquid from each plate by pipetting or scraping the top agar off with an L-spreader and pooling in an Oak Ridge centrifuge tube. It is important to use as little buffer as possible so as to not dilute the new stock.
10. Centrifuge for 30 min at 23,300 × *g* at 4 °C to remove agar pieces and bacterial debris. The phagemid stock should now be clarified.
11. Filter phage stock with a 0.45 μm filter and store at 4 °C (crude phage stock). To further purify the crude phage stock, if needed, ultracentrifuge at 82,705 × *g* for 6 h at 4 °C using a SW28 rotor. Dissolve pellet in MP buffer 200 μl and wash with 50–100 μl of 2× MP buffer (*see* **Note 6**).

12. Titer new phage stock to determine its concentration (*see* **Note 7**). This is done by plating phagemid dilutions on *M. smegmatis* cells. Ideally, the titer should be at least 1×10^{10} PFU per ml or better.

3.3 Transductions

A schematic representation of the entire transduction and collection procedure encompassing Subheadings 3.3–3.5 are depicted in Fig. 1.

1. Add 1 ml of MP buffer to the *Map* bacteria concentrated from **step 3** of Subheading 3.1.
2. Infect *Map* (1.0×10^7 CFU in 0.1 ml) with 0.1 ml of φMycoMarT7 at 1.0×10^8 PFU (*see* **Note 8**) to have a multiplicity of 10 PFU per bacterium, based on the phage titer determined at the permissive temperature for propagation (30 °C). Set up multiple aliquots if plating multiple plates (1 for each plate). Also have separate control tubes with just bacteria and phage.
3. Leave the *Map* and phagemid mixture incubating at 37 °C overnight. At the nonpermissive temperature (37 °C), the phage acts as a nonlytic transducing phage for transposon delivery into *Map*.

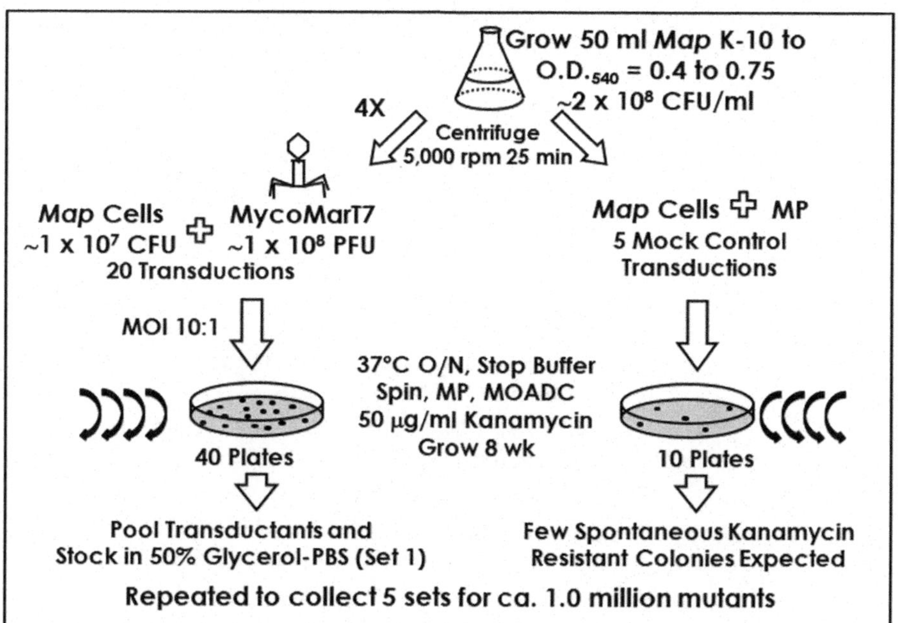

Fig. 1 Flow diagram showing the steps on how to generate a Map MycoMarT7 mutant library. The left side of the flow diagram shows the transduction sets to be pooled and stocked while the right side shows the mock transductions (without phage) to determine percentage of spontaneous kanamycin resistance

4. Add an equal volume of stop buffer, wait 5 min and immediately proceed to plating in the next section.

3.4 Plating Transposon Mutants

1. Independent transductions can be performed and plated separately to obtain a collection of insertional mutants. Alternatively, colonies can be pooled from several independent transductions to obtain a highly representative insertional library.
2. Plate 0.1 ml volumes of the stopped transduction mixture on Middlebrook 7H10 plates supplemented with 0.5% glycerol, 0.01% cycloheximide, 1 mg of mycobactin J and OADC without Tween 80, but containing either 50 μg/ml kanamycin. Use a bent glass rod "hockey stick" or L-spreader to disperse the aliquot over the entire surface of the plate (*see* **Note 4**).
3. Individual kanamycin-resistant transductants are randomly picked after an 8 week incubation period at 37 °C.
4. The entire library can be harvested and stocked in 50% glycerol-PBS freezing solution. The library can be stored in 0.5-ml aliquots at −80 °C.

3.5 Harvesting Colonies Representing Transposon Mutants

1. Colonies can be picked individually or harvested together depending on the goal of the experiment.
2. If picking and stocking individual colonies separately, these can be picked using a sterile needle into 96-well plates containing Middlebrook 7H9 media (0.2 ml).
3. If harvesting together, flood the plates with either MP buffer or Middlebrook 7H9 media and concentrate mutants by centrifugation.
4. DNA can then be extracted for Tn-seq analysis or individually stocked mutants can be screened for various phenotypic traits.

4 Notes

1. The ADC/OADC supplement is purchased commercially, filtered through a 0.2 μm filter assembly and added to the autoclaved-sterilized Middlebrook 7H9 media. Note that ADC/OADC is only added after the media has cooled below 50 °C. Bacteriological media for plating transductions should be prepared fresh 2 days before the experiment and allowed to dry in a 37 °C incubator to prevent water accumulation from condensation. Alternatively for Middlebrook 7H9 and 7H10 media preparation, ADC/OADC can be made from scratch with additional details found in Chacon et al. [20] and Larsen et al. [21]. For the growth of *M. smegmatis* mc^{2}155 and *Map* K-10, the addition of catalase and cycloheximide to these media is optional.

2. Tween™ 80 helps prevent clumping, which is not necessary when plating on solid media.
3. High titer lysates are used for propagation of phage when large amounts of phage are needed for transductions. Examples include replenishing working stocks, phage purification, or phage DNA isolation.
4. Solid media petri plates come in various sizes but the most common is 100 mm × 15 mm (60 cm^2 area) can be sealed with paraffin to prevent drying.
5. The phagemid will diffuse into the MP buffer during the overnight incubation. The plates can be kept stagnant or on a slow rocking platform. The phage soft agar will shrink at 4 °C and then expand under the 37 °C temperature to help release the phage Larsen et al. [21].
6. Keep the mycobacteriophage in two stock vials: working stock and back-up stock. Then only use from the working stock bottles and keep the back-up vial only for repropagation of phage when the working stock is depleted or has developed a problem such as contamination.
7. All phage stocks must be kept in the refrigerator at 4 °C. If the phage must be taken out of the refrigerator for any length of time, then the phage should be placed on ice. Temperature variations will cause a phage titer to decrease.
8. When using the phage stocks, extreme care must be used to prevent contamination of the stocks. We suggest wearing alcohol-rinsed latex gloves, wiping the phage stock bottle off with alcohol tissue and allowing it to dry thoroughly, using fresh sterile pipet tips every time an aliquot is taken from the phage stock bottle, and wrapping the bottle lid with paraffin before returning to 4 °C storage.

Acknowledgments

This work was supported by the USDA-Agricultural Research Service and the USDA-National Institute of Food and Agriculture grant award 2013-67015.

References

1. Foley-Thomas EM, Whipple DL, Bermudez LE, Barletta RG (1995) Phage infection, transfection and transformation of *Mycobacterium avium* complex and *Mycobacterium paratuberculosis.* Microbiology 141(Pt 5):1173–1181
2. Harris NB, Feng Z, Liu X, Cirillo SL, Cirillo JD, Barletta RG (1999) Development of a transposon mutagenesis system for *Mycobacterium avium subsp. paratuberculosis.* FEMS Microbiol Lett 175(1):21–26. doi:S0378-1097(99)00170-6 [pii]
3. Cirillo JD, Barletta RG, Bloom BR, Jacobs WR Jr (1991) A novel transposon trap for

mycobacteria: isolation and characterization of IS1096. J Bacteriol 173(24):7772–7780

4. Rathnaiah G, Lamont EA, Harris NB, Fenton RJ, Zinniel DK, Liu X, Sotos J, Feng Z, Livneh-Kol A, Shpigel NY, Czuprynski CJ, Sreevatsan S, Barletta RG (2014) Generation and screening of a comprehensive *Mycobacterium avium* subsp. *paratuberculosis* transposon mutant bank. Front Cell Infect Microbiol 4:144. https://doi.org/10.3389/fcimb.2014.00144
5. Ghosh P, Wu CW, Talaat AM (2013) Key role for the alternative sigma factor, SigH, in the intracellular life of *Mycobacterium avium subsp. paratuberculosis* during macrophage stress. Infect Immun 81(6):2242–2257. https://doi.org/10.1128/IAI.01273-12
6. Shippy DC, Lemke JJ, Berry A, Nelson K, Hines ME 2nd, Talaat AM (2017) Superior protection from live-attenuated vaccines directed against Johne's disease. Clin Vaccine Immunol 24(1). https://doi.org/10.1128/CVI.00478-16
7. Chen JW, Scaria J, Chang YF (2012) Phenotypic and transcriptomic response of auxotrophic *Mycobacterium avium subsp. paratuberculosis leuD* mutant under environmental stress. PLoS One 7(6):e37884. https://doi.org/10.1371/journal.pone.0037884
8. Park KT, Dahl JL, Bannantine JP, Barletta RG, Ahn J, Allen AJ, Hamilton MJ, Davis WC (2008) Demonstration of allelic exchange in the slow-growing bacterium *Mycobacterium avium* subsp. *paratuberculosis*, and generation of mutants with deletions at the *pknG*, *relA*, and *lsr2* loci. Appl Environ Microbiol 74(6):1687–1695. https://doi.org/10.1128/AEM.01208-07. AEM.01208-07 [pii]
9. Park KT, Allen AJ, Barrington GM, Davis WC (2014) Deletion of *relA* abrogates the capacity of *Mycobacterium avium paratuberculosis* to establish an infection in calves. Front Cell Infect Microbiol 4:64. https://doi.org/10.3389/fcimb.2014.00064
10. Faisal SM, Chen JW, Yan F, Chen TT, Useh NM, Yan W, Guo S, Wang SJ, Glaser AL, McDonough SP, Singh B, Davis WC, Akey BL, Chang YF (2013) Evaluation of a *Mycobacterium avium* subsp. *paratuberculosis leuD* mutant as a vaccine candidate against challenge in a caprine model. Clin Vaccine Immunol 20(4):572–581. https://doi.org/10.1128/CVI.00653-12
11. Rathnaiah G, Bannantine JP, Bayles DO, Zinniel DK, Stabel JR, Grohn YT, Barletta RG (2016) Analysis of *Mycobacterium avium subsp. paratuberculosis* mutant libraries reveals loci-dependent transposition biases and strategies to novel mutant discovery. Microbiology. https://doi.org/10.1099/mic.0.000258
12. Danelishvili L, Wu M, Stang B, Harriff M, Cirillo SL, Cirillo JD, Bildfell R, Arbogast B, Bermudez LE (2007) Identification of *Mycobacterium avium* pathogenicity island important for macrophage and amoeba infection. Proc Natl Acad Sci U S A 104(26):11038–11043. https://doi.org/10.1073/pnas.0610746104. 0610746104 [pii]
13. Wu CW, Livesey M, Schmoller SK, Manning EJ, Steinberg H, Davis WC, Hamilton MJ, Talaat AM (2007) Invasion and persistence of *Mycobacterium avium* subsp. *paratuberculosis* during early stages of Johne's disease in calves. Infect Immun 75(5):2110–2119. https://doi.org/10.1128/IAI.01739-06. IAI.01739-06 [pii]
14. Bannantine JP, Everman JL, Rose SJ, Babrak L, Katani R, Barletta RG, Talaat AM, Grohn YT, Chang YF, Kapur V, Bermudez LE (2014) Evaluation of eight live attenuated vaccine candidates for protection against challenge with virulent *Mycobacterium avium* subspecies *paratuberculosis* in mice. Front Cell Infect Microbiol 4:88. https://doi.org/10.3389/fcimb.2014.00088
15. Shin SJ, Wu CW, Steinberg H, Talaat AM (2006) Identification of novel virulence determinants in *Mycobacterium paratuberculosis* by screening a library of insertional mutants. Infect Immun 74(7):3825–3833. https://doi.org/10.1128/IAI.01742-05. 74/7/3825 [pii]
16. Bardarov S, Kriakov J, Carriere C, Yu S, Vaamonde C, McAdam RA, Bloom BR, Hatfull GF, Jacobs WR Jr (1997) Conditionally replicating mycobacteriophages: a system for transposon delivery to *Mycobacterium tuberculosis*. Proc Natl Acad Sci U S A 94(20):10961–10966
17. Sassetti CM, Boyd DH, Rubin EJ (2001) Comprehensive identification of conditionally essential genes in mycobacteria. Proc Natl Acad Sci U S A 98(22):12712–12717. https://doi.org/10.1073/pnas.231275498
18. Rubin EJ, Akerley BJ, Novik VN, Lampe DJ, Husson RN, Mekalanos JJ (1999) In vivo transposition of mariner-based elements in enteric bacteria and mycobacteria. Proc Natl Acad Sci U S A 96(4):1645–1650
19. Reddacliff LA, Nicholls PJ, Vadali A, Whittington RJ (2003) Use of growth indices from radiometric culture for quantification of sheep strains of *Mycobacterium avium* subsp. *paratuberculosis*. Appl Environ Microbiol 69(6):3510–3516

20. Chacon O, Bermudez LE, Zinniel DK, Chahal HK, Fenton RJ, Feng Z, Hanford K, Adams LG, Barletta RG (2009) Impairment of D-alanine biosynthesis in *Mycobacterium smegmatis* determines decreased intracellular survival in human macrophages. Microbiology 155(Pt 5):1440–1450. https://doi.org/10.1099/mic.0.024901-0

21. Larsen MH, Biermann K, Jacobs WR Jr (2007) Laboratory maintenance of *Mycobacterium tuberculosis*. Curr Protoc Microbiol. Chapter 10:Unit 10A 11. https://doi.org/10.1002/9780471729259.mc10a01s6

Part II

Transposon General Protocols and Applications

Chapter 12

Protocols of Conjugative Plasmid Transfer in *Salmonella*: Plate, Broth, and Filter Mating Approaches

Bijay K. Khajanchi, Pravin R. Kaldhone, and Steven L. Foley

Abstract

Bacterial conjugation is a natural process that allows for horizontal transmission of DNA from one bacterium to another. Several plasmids carry transposons that encode multiple antimicrobial and metal resistance genes. Conjugative plasmid transfer requires intimate cell-to-cell contacts between the donor and the recipient. Self-conjugative plasmids harbor *tra* genes which facilitate plasmid transfer from donor to recipient bacterial strain. Here we describe different methods of conjugative plasmid transfers via conjugation.

Key words Conjugation, Plasmid, *Salmonella*, *tra* genes, Transposons

1 Introduction

Bacterial conjugation is a natural mechanism employed by many bacterial species to allow for the horizontal transmission of DNA from one bacterium to another through direct cell-to-cell contact. In Gram-negative bacteria, complex extracellular filaments such as sex pili facilitate physical contacts between donor and recipient bacteria. The DNA macromolecules that are typically transferred are plasmids; however, conjugative transposons can be transferred. DNA transfer via conjugation is often termed horizontal gene transfer (HGT). Plasmids in *Salmonella enterica* are extrachromosomal, circular DNA molecules that often contain genes that provide the bacterial host with additional characteristics such as virulence or antimicrobial resistance [1–3]. Conjugative plasmid transfer requires intimate cell-to-cell contacts between donor and recipient bacterial cells. The recipient bacteria that receive plasmid (s) from donor cells are known as transconjugants. Transposon mutagenesis can be used to inactivate genes on plasmids allowing for detailed evaluations of plasmid function. These mutated plasmids can then be transferred via conjugation into recipient cells and the transconjugants assessed to determine the impact of the

Steven C. Ricke et al. (eds.), *Microbial Transposon Mutagenesis: Protocols and Applications*, Methods in Molecular Biology, vol. 2016, https://doi.org/10.1007/978-1-4939-9570-7_12, © Springer Science+Business Media, LLC, part of Springer Nature 2019

mutation on bacterial host physiology through comparisons to the wild-type plasmid carrying strains. Additionally, plasmids are often used to complement genes knocked out in the host chromosome back into the mutant host to determine if the reintroduction of the gene leads to a return of function.

Within the plasmids themselves, genes can be acquired through various HGT mechanisms including integration of transposons, phages, integrons, or other insertion sequences. Several resistance plasmids have been shown to carry transposons that encode multiple metal and antimicrobial resistance genes. Many plasmids, including those that carry transposons, are self-transmissible through the process of conjugation. The conjugation machinery is encoded by transfer (*tra*) genes that allow the plasmids to be transferred to a recipient strain, potentially increasing antimicrobial resistance and virulence of the recipients [4]. Bacterial cells can carry multiple plasmids. To retain multiple plasmids in the same bacterial cell, these plasmids cannot be from the same incompatibility group, which encodes identical replication or partition systems, a phenomenon known as plasmid incompatibility [5]. Figure 1 shows a schematic diagram of incompatibility group A/C (IncA/C) plasmid of *S.* Heidelberg.

During conjugation, a conjugative pilus is formed that consists of several pilin subunits encoded by multiple *tra* genes such as *traA*, *traQ*, and *traX* [6]. Once conjugative pilus has bridged between donor and recipient bacteria, nucleoprotein complex formation initiated at the *ori*T (origin of transfer) on the F factor. F-plasmid-encoded proteins TraI and TraY facilitate the formation of a nucleoprotein complex [7, 8], and several other factors contribute to the process to facilitate the transfer of the plasmid DNA from the donor to the recipient [9, 10].

The transferability of plasmids allows researchers to assess various biological functions of genes carried on the plasmids within a naïve background by generating transconjugants. Because many plasmids carry antimicrobial resistance markers, approaches can be designed to transfer plasmids of interest from the donor strain to a well characterized recipient through the process of coselection [11]. A mixture of the plasmid-containing donor and the recipient carrying a chromosomally encoded resistance marker (that the donor lacks) grown in the presence of both the antimicrobials whose resistance is encoded on the plasmid and chromosomally encoded on the recipient will only select for the transconjugant. The contribution of the plasmid encoded factors including antimicrobial resistance and virulence can now be assessed in the new host allowing for characterization of the plasmid's contribution to a bacterium's phenotype. The following sections outline the material requirements and approaches to conduct conjugation experiments.

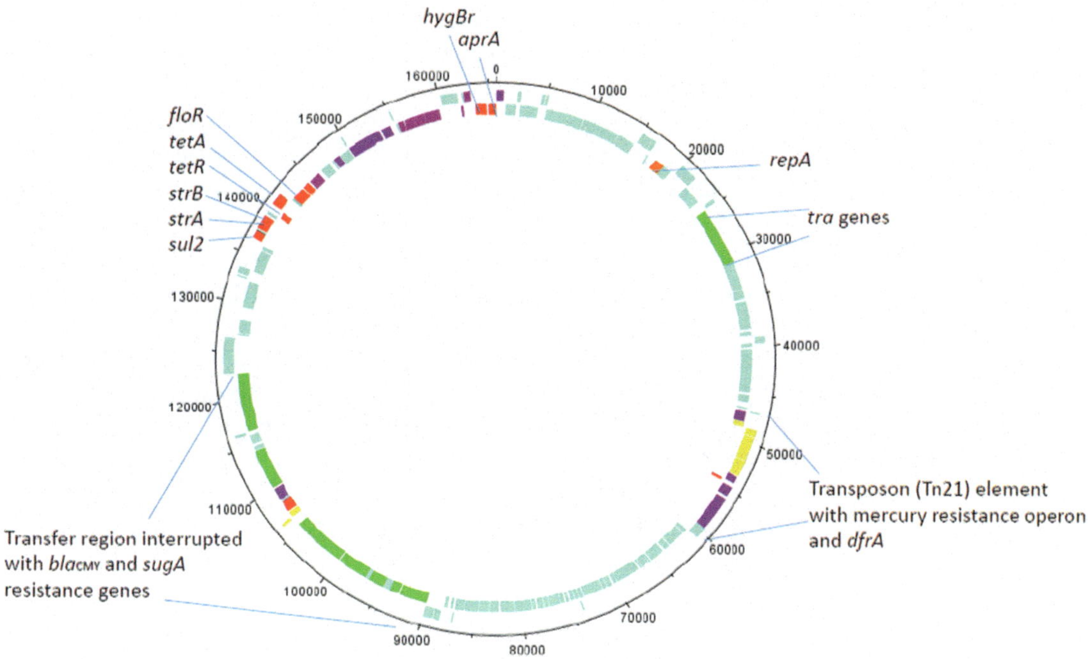

Fig. 1 Schematic diagram of IncA/C plasmid of *S.* Heidelberg isolated from cattle. This particular IncA/C plasmid possesses transfer regions, including a region interrupted with resistance genes, and transposon element (Tn21) which carries genes associated with mercury resistance and trimethoprim antimicrobial resistance (*dfrA*) [11]

2 Materials

Media will need to be prepared to grow donor and recipient strains and select for transconjugants following plasmid transfer. Depending on the experimental approach used (*see* below), nonselective and selective broth and/or agar plates containing antimicrobial agents will be required to perform the conjugation experiments. All solutions should be made using distilled deionized water (18 MΩ cm). Individual media and antimicrobial selection reagents are prepared as described below.

2.1 Nonselective Nutrient Media

1. Luria–Bertani (LB) broth: This medium contains three major ingredients, tryptone (10 g/L), yeast extract (5 g/L), and sodium chloride (5 g/L; *see* **Note 1**), and can be purchased already made, as a premixed powder from multiple suppliers or prepared from the individual components [12]. For the premixed powder, weigh out the appropriate amount of dried medium (e.g., 20 g of LB broth/L distilled water) and add to an autoclavable bottle with the desired volume of water. Heat the mixture on a magnetic stirring hotplate or in the microwave to dissolve the powder. Autoclave the liquid media at 121 °C, 15 psi for a minimum of 20 min.

2. Tryptic soy broth: This medium consists of five major ingredients: pancreatic digest of casein (17 g/L), papaic digest of soybean (3 g/L), dextrose (2.5 g/L), sodium chloride (5 g/L), and dipotassium phosphate (2.5 g/L). Dissolve the premixed powder as per manufacturer's instruction; for example, dissolve 30 g of premixed powder/L water using a magnetic stirring hot plate or heating in a microwave. Then autoclave the medium at 121 °C, 15 psi for a minimum of 20 min.
3. LB Agar: Add 20 g LB broth powder and 10 g agar to 1 L water and heat the mixture until the agar and media are fully dissolved as described above. Autoclave the solution at 121 °C for 20 min. Following autoclaving, move the media bottle to a 55 °C water bath to temper the agar. Once the temperature has equilibrated, pour the media in to 15 mm petri dishes under aseptic conditions and allow them to solidify (approximately 30 min). For poured plates, it is important to test the sterility of the plates by incubating them overnight at 37 °C prior to storage at 4 °C until use. Alternatively, premade, quality control-tested plates can be purchased from different manufacturers and used for experiments.

2.2 Stock Solution of Antibiotic and/or Other Selective Agents

1. Prepare stock solution of antibiotics such as ampicillin, gentamicin, streptomycin, tetracycline, and nalidixic acid at a concentration of 50 mg/mL by dissolving in sterile water or other suitable solvent (e.g., tetracycline is dissolved in 95% ethanol). The antimicrobial powder is weighed out and added to appropriate volume of diluent (e.g., 0.5 g of ampicillin sulfate in 10 mL of sterile water). For some of the antibiotics, the pH of the stock solution needs to be adjusted to dissolve the antimicrobial agent. Once in solution, filter solutions using 0.2 μm size filter (Millipore Corp., Burlington, MA) to remove potential microbial contamination. The antibiotic stock solutions can be stored at −20 °C.
2. Sodium azide (NaZ): A common recipient in many of conjugation experiments is *Escherichia coli* J53, which is resistant to NaZ [13] and provides an attractive marker to screen transconjugants. To prepare the NaZ stock solution, dissolve 1 g NaZ powder in 10 mL of sterile water to achieve stock concentration of 100 mg/mL. The solution should be filter-sterilized and prepared fresh before each experiment.

2.3 Selective Media

1. MacConkey Agar: Dissolve 50 g MacConkey dehydrated medium (Beckton, Dickinson and Company, Sparks, MD) in 1 L water. The major components of this agar are pancreatic digest of gelatin (17 g/L), peptones (3 g/L), lactose (10 g/L), bile salts (1.5 g/L), and sodium chloride (5 g/L). Autoclave the media at 121 °C for 20 min, cool at 55 °C to temper. Then

pour the media into petri dishes under aseptic conditions. Let the plates solidify and incubate overnight at 37 °C to assess sterility and store at 4 °C.

2. Selection plates made with either LB or MacConkey agar: LB and MacConkey agar were prepared as described before in the Subheading 2.1, **item 3** and Subheading 2.3, **item 1**, respectively. After cooling the solution at 55 °C for 30 min in water bath, add the respective amount of antimicrobial to achieve following final concentration of selective agents (*see* **Note 2**) (e.g., ampicillin (32 μg/mL), gentamicin (10 μg/mL), tetracycline (32 μg/mL), or streptomycin (32 μg/mL) and NaZ (350 μg/mL) or nalidixic acid (64 μg/mL)). Then, mix the solution by gentle stirring to disperse the antibiotic (s) thoroughly throughout the media. Pour the media into petri dishes using aseptic techniques as noted above. Let the plates solidify, incubate overnight at 37 °C to assess sterility and store at 4 °C. It is advisable to streak donor and recipient strains on selection plates used for transconjugants screening to make sure they do not grow in monoculture. This helps ensure the efficacy of the selection process for transconjugants.

2.4 Suitable Recipient(s)

In preparation for conjugation experiments, the selection of an appropriate recipient is vital for several reasons. First, the recipient must have a selectable marker, such as an antimicrobial resistance phenotype, that the plasmid-carrying donor lacks to allow for selection of successful transconjugants. Additionally, if trying to determine the contribution of a particular phenotype encoded on the plasmid in a naïve background, the recipient must not express the phenotype, or express it at a low enough level, so that the contribution of the plasmid to the phenotype can be determined. To illustrate these points, in a recent study in our laboratory, a two-step approach was used to evaluate the contributions of plasmids to survival of *Salmonella* in intestinal epithelial cells [14]. Based on the antimicrobial resistance properties of the donor *Salmonella* strains, transfer of the plasmids directly to a nalidixic acid or rifampicin-resistant *Salmonella* recipient was not feasible for the selection of transconjugants. Thus, the first step was to transfer the plasmids from the *Salmonella* strains to *E. coli* J53, a NaZ-resistant strain [13] that was susceptible to nalidixic acid. The second step, was to transfer the plasmids of interest from the *E. coli* J53 transconjugants into a separate *Salmonella* strain lacking the plasmid (*see* **Note 3**). This approach allowed for effective assessment of the contribution of the plasmids to invasion and persistence of *Salmonella* in intestinal cells [14].

2.5 Development of Resistance Phenotype in Recipient

In specific conjugation experiments, it is necessary to develop a phenotypic marker on recipient to select successful transconjugants. In our experiments, we have generated and used nalidixic acid and/or rifampicin-resistant mutant *Salmonella* strains as recipients because they had other desired characteristics needed for our studies. To develop the mutants, the strains were serially passaged on selective media containing gradually increasing concentration of nalidixic acid or rifampicin (2–128 μg/mL). As the bacterial cells grow on the selective media, representative colonies were selected and streaked onto selection plates with double the concentration of the antibiotic. This approach selected for bacteria that had spontaneous mutations in chromosomally encoded genes, such as *gyrA*, that convey resistance to nalidixic acid. Once the minimal inhibitory concentration (MIC) reaches a level sufficient for selection in conjugation experiments (e.g., 64–128 μg/mL), the strain can be frozen in a freezing medium (e.g., brain–heart infusion broth with 20% glycerol) at −80 °C for long-term storage until needed for conjugation experiments.

3 Methods

3.1 Plate Mating Conjugation Approach

1. Streak donors and the recipient separately on LB agar plates and on the selection plates.
2. Incubate plates overnight at 37 °C.
3. After incubation, ensure that there is no growth on the selection plates.
4. From the LB plates, use a sterile loop to pick a colony from the recipient plate and streak in a single line across an LB agar plate, next pick a colony from the donor plate and streak across where you just streaked the recipient. Multiple recipient–donor crosses can be made on a single plate.
5. Incubate plates overnight at 37 °C.
6. The following day, at the point of intersection, pick the bacterial cells equivalent of three to four colonies and streak onto selection plates containing appropriate selection agents (*see* **Note 2**).
7. Incubate the selection agar plates up to 48 h at 37 °C to observe for growth.
8. If any growth is seen on selection plates, pick representative colonies and restreak them on selection plates and incubate overnight at 37 °C.
9. Colonies that grow on these plates should be evaluated to confirm the identity of the transconjugants (*see* below) and prepared for archival freezing at −80 °C [15].

3.2 Broth Mating Conjugation Approach

1. Prepare the strains by inoculating the donors and the recipient isolates separately in 2 mL LB broth.

2. Incubate the cultures overnight at 37 °C with shaking at 180 revolutions per minute.
3. For each mating experiment, combine 100 μL of the donor and 100 μL of the recipient cultures (1:1 ratio, *see* **Note 4**) in a 1.5 mL tube.
4. Briefly vortex the mixed culture for a few seconds.
5. Incubate the cultures at 37 °C for 4 h.
6. Spread 100 μL of the cell suspension on appropriate selection agar plates (*see* **Note 2**).
7. Incubate the selection plates up to 48 h at 37 °C.
8. If any growth is seen on selection plates, pick representative colonies and restreak them on selection plates and incubate overnight at 37 °C. Colonies that grow on these plates should be evaluated to confirm the identity of the transconjugants (*see* below) and prepared for archival freezing at −80 °C [15].

3.3 Filter Mating Conjugation Approach

1. Prepare overnight broth cultures of donor and recipient cells as described above.
2. Prepare filter apparatuses by placing a 0.45 mm membrane filter in a stainless-steel filter holder (Millipore Corp., Burlington, MA) and sterilize each filter apparatus by autoclaving it in an instrument pouch.
3. For each mating experiment, combine 100 μL each of the donor and recipient cultures (1:1 ratio, *see* **Note 4**) and add 800 μL of fresh LB broth.
4. Mix cell suspension using vortex mixer for few seconds and draw the suspension up into a 3 cm^3 or 5 cm^3 luer-lock syringe.
5. Remove a sterilized filter apparatus from its pouch and ensure that it is tightly screwed together to avoid leakage. Connect the syringe containing the culture to the luer-lock connection and pass the culture through filter.
6. Properly discard the flow-through media and unscrew the stainless-steel filter housing, aseptically remove the filter with a sterilized forceps and place the filter on sheep's blood agar plates (purchased from a commercial vendor) and incubate at 37 °C overnight.
7. The following day, with a sterile inoculating loop, pick bacterial growth adjacent to the filter and suspend in 1 mL of LB broth.
8. Vortex the mixture and spread 100 μL of the cell suspension on appropriate selection agar plates (*see* **Note 2**) and proceed with the selection process as described for the broth mating experiments.

3.4 Confirmation of Transconjugants

1. There are multiple approaches used to confirm the identity of transconjugants, here we briefly mention overall approaches, with references for the detailed methods to conduct the testing. For each of these methods, there is a requirement that relevant information on the genotypes and phenotypes of the donor and recipient strains are known.
2. Growth on selective media: often plasmids from *Salmonella* are transferred into *E. coli* and when the transconjugants are grown on MacConkey agar with antibiotics that select for the plasmids, most *E. coli* strains ferment lactose and grow as pink colonies, which serves as an indication that transconjugants were formed.
3. Direct isolation and separation of plasmids can be used to confirm the transfer of the plasmid to the recipient. Plasmid DNA can be isolated using commercial kits such as the Qiagen Miniprep kit (Qiagen Inc., Valencia CA) following the manufacturer's instructions. Isolated plasmids can then be separated, along with supercoiled DNA ladders in 0.7% agarose gels, stained, and visualized under UV light. The plasmids present in the putative transconjugant are compared to the donor strains to assess which plasmid(s) transferred [16].
4. Plasmids can also be characterized using PCR-based replicon typing methods [17] to predict the plasmid incompatibility (Inc) groups of the plasmids in the strains. It is important to use proper positive controls for these experiments as described by Carattoli et al. (2005) for the PCR reactions [18]. The results of the PCR reactions for the donors, the recipient, and the transconjugants are compared to identify the plasmids that transferred. Please note that not all replicon types are represented in the PCR-based typing scheme.
5. Because the plasmids that are transferred contain antimicrobial resistance genes, antimicrobial susceptibility testing (AST) can be used to characterize the transconjugants in relationship to those of the donor and the recipient. The AST is conducted by disk diffusion or broth microdilution testing following Clinical and Laboratory Standard Institute (CLSI) guidelines [19, 20]. The test panel should include antimicrobials relevant to the donor and recipient phenotypes to evaluate the conjugation experiments. Quality control strains such as *E. coli* ATCC 25922 and *Staphylococcus aureus* ATCC 29213 should be included in the testing regimen.
6. Whole genome sequencing of the transconjugants can also be used to confirm the identity of transconjugants. Khajanchi et al. (2016) describes methods for whole genome sequencing of *Salmonella* strains [21].

4 Notes

1. There are multiple different formulations of "LB" media, including LB Base, Lennox, and Miller, the main difference between the different formulations is the amount of sodium chloride present in the formulation [12]. Because of these differences, it is important to read the media bottle to ensure that the proper amount of medium is dissolved in the solution.
2. It is important to know that antimicrobial resistance phenotypes of the donors and potential recipients to develop a selection strategy for the transconjugants. At a minimum, each strain in the study should undergo antimicrobial susceptibility testing. Many of the genes encoding antimicrobial resistance in *Salmonella* are known to be plasmid encoded but not all. Additionally, many *Salmonella* strains contain multiple resistance plasmids, so the genes encoding the resistance may not be on the plasmid of interest. The use of whole genome sequencing of the donor strains can help to identify the resistance genes present and provide information on the locations of the genes. Information on the resistance genes and their locations will allow you to choose the appropriate selection antimicrobial to screen for transconjugants with the plasmid of interest. For example, in our experiments, antimicrobial resistance phenotypes and corresponding plasmid-encoded genes from donors led to the use of ampicillin (32 μg/mL), gentamicin (16 μg/mL), streptomycin (32 μg/mL), or tetracycline (32 μg/mL) for selection of plasmid transfer. In addition, chromosomally encoded resistance markers in the recipient, that the donor lacked, such as NaZ (350 μg/mL) or nalidixic acid (128 μg/mL), are also added to the selection plates for the transconjugants.
3. Since *Salmonella* commonly carry multiple different plasmids within the same bacterial cell, it is sometimes important to obtain a transconjugant that contain only a single targeted plasmid in *Salmonella*. To accomplish this, we established a two-step approach [14], in which the plasmids present in the wild-type *S. enterica* strain such as SE163A (*see* **Note 5**) are transferred to the NaZ resistant recipient *E. coli* J53 using a plate mating strategy. Transconjugants on *E. coli* J53 background are screened for the plasmids that the donor harbored using PCR-based replicon typing as described above. In the second step, transconjugants that have the target plasmid but lack others are used as donors for the transfer to a *Salmonella enterica* recipient, such as SE819 described in the paper, that lacks target plasmid but has a selection marker using a slightly different approach [14]. Briefly, a single colony of

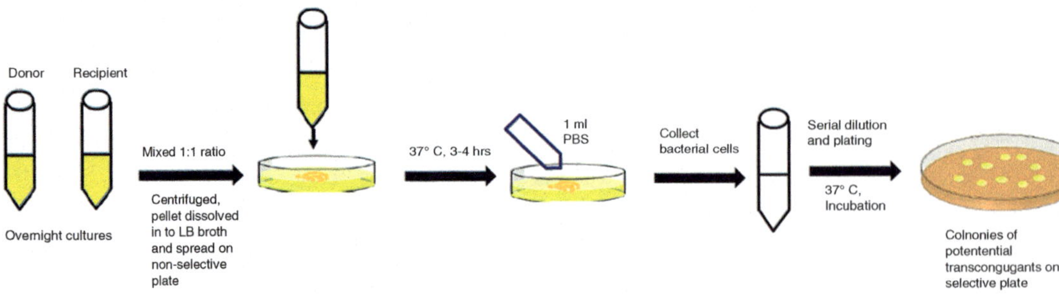

Fig. 2 Plasmid transfer from *E. coli* J53 (donor) to *S. enterica* SE819 (recipient)

transconjugant on *E. coli* J53 background (donor) and the recipient SE819 are grown separately in LB broth overnight. The recipient and the donor are subsequently mixed together in a 1:1 proportion and centrifuged at 13,000 × *g* for 5 min to obtain the pellet. The pellets are dispersed in 250 μL of LB broth and spotted onto LB agar plate. The plate is incubated for 3–4 h at 37 °C in upright position. The growth seen is dissolved in 1 mL PBS and 100 μL of cell suspension is spread onto MacConkey agar plate containing selective antibiotics. After 18–20 h incubation, individual colorless colonies are picked, grow overnight on selective LB broth, and frozen for further analyses (Fig. 2).

4. In broth mating assay, the 1:1 donor–recipient ratio gave us optimal efficiency. In our previous study, we started with 1:9 ratio and noticed that it did not work for some of the conjugation experiments. In such cases, increasing donor–recipient ratio up to 1:1 yielded successful conjugation [15]. It may be valuable to try multiple different donor–recipient ratios to optimize plasmid transfer efficiency.
5. To evaluate the antimicrobial resistances, plasmid contents and genetic contents of donor SE163A and recipient SE819 strains, these strains were sequenced and deposited into GenBank under the accession numbers LSZD00000000 and LSZE00000000, respectively [21].

Acknowledgments

The authors would like to thank Drs. Ashraf Khan and Jing Han for their insightful review of the book chapter. The opinions expressed in this book chapter are solely the responsibility of the authors and do not necessarily represent the official views and policy of the US Food and Drug Administration. Reference to any commercial materials, equipment, or process does not in any way constitute approval, endorsement, or recommendation by the Food and Drug Administration.

References

1. Richter-Dahlfors A, Buchan AM, Finlay BB (1997) Murine salmonellosis studied by confocal microscopy: *Salmonella typhimurium* resides intracellularly inside macrophages and exerts a cytotoxic effect on phagocytes in vivo. J Exp Med 186:569–580
2. Rotger R, Casadesus J (1999) The virulence plasmids of *Salmonella*. Int Microbiol 2:177–184
3. Sheppard M, Webb C, Heath F et al (2003) Dynamics of bacterial growth and distribution within the liver during *Salmonella* infection. Cell Microbiol 5:593–600
4. Ahmer BM, Tran M, Heffron F (1999) The virulence plasmid of *Salmonella typhimurium* is self-transmissible. J Bacteriol 18:1364–1368
5. Novick RP (1987) Plasmid incompatibility. Microbiol Rev 51:381–395
6. Silverman PM (1997) Towards a structural biology of bacterial conjugation. Mol Microbiol 23:423–429
7. Nelson W, Howard M, Sherman J et al (1995) The traY gene product and integration host factor stimulate *Escherichia coli* DNA helicase I-catalyzed nicking at the F plasmid *oriT*. J Biol Chem 270(28):374–328, 380
8. Howard M, Nelson W, Matson S (1995) Stepwise assembly of a relaxosome at the F plasmid origin of transfer. J Biol Chem 270:28381–28386
9. Matson SW, Sampson JK, Byrd DRN (2001) F plasmid conjugative DNA transfer. J Biol Chem 276:2372–2379
10. Tsai MM, Fu YH, Deonier RC (1990) Intrinsic bends and integration host factor binding at F plasmid *oriT*. J Bacteriol 172:4603–4609
11. Han J, Lynne AM, David DE et al (2012) DNA sequence analysis of multidrug resistance encoding plasmids from *Salmonella enterica* serotype Heidelberg isolates. PLoS One 7: e51160
12. Zimbro MJ, Power DA (2003) Difco & BBL Manual: Manual of Microbiological Culture Media. Becton, Dickinson and Company, Sparks, MD
13. Jacoby GA, Han P (1996) Detection of extended-spectrum betalactamases in clinical isolates of *Klebsiella pneumoniae* and *Escherichia coli*. J Clin Microbiol 34:908–911
14. Khajanchi BK, Hasan NA, Choi SY et al (2017) Comparative genomic analysis and characterization of incompatibility group FIB plasmid encoded virulence factors of *Salmonella enterica* isolated from food sources. BMC Genomics 18:570. https://doi.org/10.1186/s12864-017-3954-5
15. Kaldhone PR, Han J, Deck J et al (2018) Evaluation of the genetics and functionality of plasmids in Incompatibility Group I1 (IncI1) positive *Salmonella enterica*. Foodborne Pathog Dis 15(3):168–176. https://doi.org/10.1089/fpd.2017.2332
16. Kaldhone PR, Nayak R, Lynne AM et al (2008) Characterization of *Salmonella enterica* serovar Heidelberg from turkey-associated sources. Appl Environ Microbiol 74:5038–5046
17. Johnson TJ, Nolan LK (2009) Plasmid replicon typing. Methods Mol Biol 551:27–35
18. Carattoli A, Bertini A, Villa L et al (2005) Identification of plasmids by PCR-based replicon typing. J Microbiol Methods 63:219–228
19. CLSI (2018a) Performance standards for antimicrobial disk and dilution susceptibility tests for bacteria isolated from animals (VET01), 5th edn. Clinical and Laboratory Standards Institute, Wayne, PA
20. CLSI (2018b) Performance standards for antimicrobial susceptibility testing (M100), 28th edn. Clinical and Laboratory Standards Institute, Wayne, PA
21. Khajanchi BK, Han J, Gokulan K et al (2016) Draft genome sequences of four *Salmonella enterica* strains isolated from turkey-associated sources. Genome Announc 4(5): e01122–e01116

Chapter 13

Construction of DNA Barcode-Tagged *Salmonella* Strains

Yichao Yang, Reena Chandrashekar, Steven C. Ricke, and Young Min Kwon

Abstract

This chapter provides a detailed protocol for construction of DNA barcode-tagged isogenic strains of *Salmonella*. The protocol is illustrated with *S.* Enteritidis in this chapter. However, this protocol should be widely applicable to other *Salmonella* serotypes. A series of the DNA barcode-tagged strains thus constructed can be used in combination with next generation sequencing or quantitative PCR to study the population dynamics of the bacterial pathogen during infection within the host or transmission within a population of the host in a quantitative manner.

Key words *Salmonella*, Population dynamics, DNA barcode-tagging, Overlapping extension PCR, Red recombination system, Next generation sequencing

1 Introduction

Studying population dynamics of a bacterial pathogen within the host or a population of the host in a quantitative way is important for better understanding of the mode of infection and transmission of the pathogen. Traditionally, marker genes conferring antibiotic resistance or encoding fluorescent proteins have been used to construct marker strains, which then can be used to track and quantify the strains in the given experimental settings. However, these approaches are limited in the number of the distinctive strains that can be studied simultaneously, and those genetic changes can alter the fitness of the wild type strains. In recent years, a new approach has been developed in which the bacterial strains are modified to carry distinctive DNA barcodes inserted into a neutral genomic locus and used in combination with high-throughput DNA sequencing to create quantitative profiles. This new barcode-tagging approach has been used to understand population dynamics of diverse bacterial or even viral species during infection in the host [1–3]. However, we have constructed a series of *S.* Enteritidis strains tagged with distinctive DNA barcodes and

Steven C. Ricke et al. (eds.), *Microbial Transposon Mutagenesis: Protocols and Applications*, Methods in Molecular Biology, vol. 2016, https://doi.org/10.1007/978-1-4939-9570-7_13,

used them to investigate their transmission dynamics within a group of chickens [4] and the competitive exclusion during chicken infection [5]. Transposon mutants with insertions in functionally neutral genomic regions might be used for the similar purpose, but the use of barcode-tagged strains provide a more versatile and reliable method to track and monitor the population dynamics of bacteria.

In this chapter, we provide a step-by-step protocol for the construction of DNA barcode-tagged strains of *S.* Enteritidis, which can be used in the similar manner for other *Salmonella* serotypes.

2 Materials

2.1 Plasmids and Oligonucleotides

1. Plasmid pKD4, which is used as a template DNA to amplify Kanamycin resistance (Km^R) gene [6] (*see* **Note 1**).
2. Plasmid pKD46, which expresses Red recombinase system under the control of arabinose-inducible promoter. pKD46 has a temperature-sensitive replicon requiring 30 °C for replication [6].
3. Oligonucleotides to be used as PCR primers (Table 1). The design and annealing sites of the primers are shown in Fig. 1.

2.2 PCR Amplification

1. Pfu DNA polymerase (2.5 U/μl) and 10× cloned Pfu DNA polymerase buffer (Stratagene).
2. TaKaRa Ex Taq polymerase (5 units/μl) and 10× Ex Taq Buffer (TaKaRa).
3. GoTaq® G2 green master mix (Promega).
4. dNTPs mixture (2.5 mM each).
5. 20 μM of each primer.
6. DNA engine thermal cycler (Bio-Rad).

2.3 DNA Extraction and Analysis

1. Agarose gel electrophoresis system.
2. QIAamp DNA Mini Kit (Qiagen).
3. QIAquick gel extraction kit (Qiagen).
4. Qubit fluorometer and Qubit dsDNA HS assay kit (Invitrogen).

2.4 Electroporation

1. Luria–Bertani (LB) broth.
2. Kanamycin (Km) 50 μg/ml and ampicillin (Amp) 100 μg/ml (final concentrations).
3. 10 mM L-arabinose (final concentration).

Table 1
List of oligonucleotides

Name	DNA sequence (5′ → 3′)
Up-F	GCAAGGTTGGTGTCTGTCCT
Up-R-BC-P1	GAAGCAGCTCCAGCCTACACNNNNNNATTATTGTTAATTTATTCTT[a]
P1	GTGTAGGCTGGAGCTGCTTC
P2	ATGGGAATTAGCCATGGTCC
Dn-F-P2	GGACCATGGCTAATTCCCATAAAGGTTAAGCAGTGACCCA[a]
Dn-R	GTTGATGGACTGGGTTCGTT
BC-F	AGCGTCCTGAAATAATAAAAGAA
BC-R	CGGACTGGCTTTCTACGTGT

[a]Underlined sequences are reverse complementary to P1 or P2 primer sequences

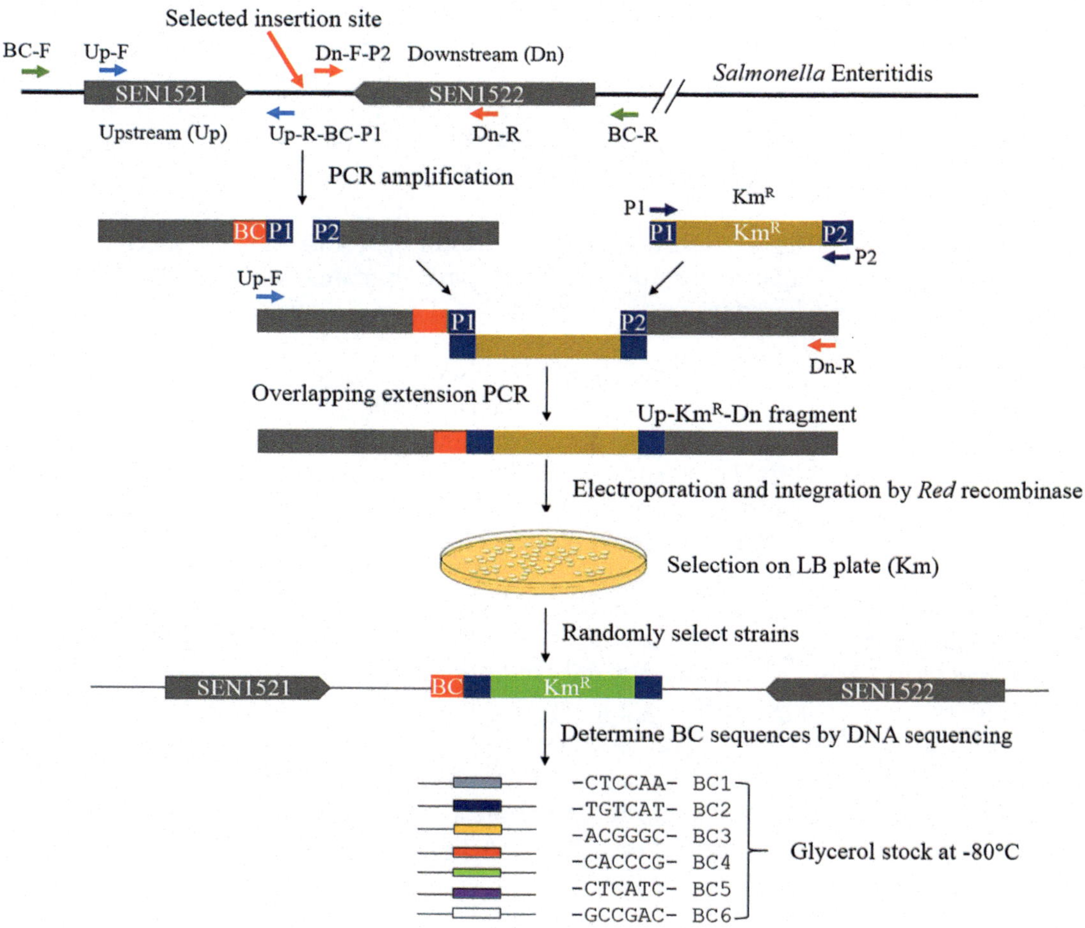

Fig. 1 Construction of barcode-tagged strains of *Salmonella*

4. ECM 399 exponential decay wave electroporation system (BTX).
5. Super Optimal Catabolite (SOC) media.

3 Methods

3.1 Overview

The entire process largely consists of the three following steps as shown in Fig. 1. In the first step, the upstream (Up) and downstream (Dn) fragments of the selected insertion site (the intergenic region between SEN1521 and SEN1522 in this illustration) and the Kanamycin resistance (Km^R) gene are amplified by PCR. The PCR primers are designed so that the resulting fragments have ~20 bp overlapping regions between the two fragments to be joined together in the next step. In the illustration shown in Fig. 1, the overlapping regions are P1 between Up and Km^R fragments, and P2 between Km^R and Dn fragments. One of those primers is also designed to include the DNA barcode of random nucleotides (BC) to the 3′ of the overlapping region (Up-R-BC-P1 in Fig. 1). In the second step, the 3 fragments (Up, Km^R, and Dn) are joined together via overlapping extension PCR. In the third step, the joined fragment (Up-Km^R-Dn) is transformed into *Salmonella* cells expressing Red recombinase system to allow efficient homologous recombination of the linear DNA fragment into the target insertion site. The resulting transformants selected on LB agar plate supplemented with Kanamycin carry distinct DNA barcodes.

3.2 Selection of Functionally Neutral Genomic Locus

DNA barcode should be inserted into the genomic region where the insertion does not change the phenotype of the resulting strains. In our previous studies, the intergenic region between SEN1521 and SEN1522 was selected as a functionally neutral genomic locus in *S.* Enteritidis and used as the insertion site for DNA barcodes [4, 5] (*see* **Note 2**).

3.3 Preparation of 3 Individual Fragments: Up, Km^R and Dn

1. Extract genomic DNA from *S.* Enteritidis strain using QIAamp DNA Mini Kit. The genomic DNA is used as the template to amplify Up and Dn fragments.
2. PCR amplify Up fragment (using Up-F and Up-R-BC-P1 primers) and Dn fragment (using Dn-F and Dn-R) through 30 cycles of 94 °C for 30 s, 55 °C for 60 s, 72 °C for 1 min followed by extension at 72 °C for 10 min using the PCR reaction mix shown in Table 2 (*see* **Note 3**).
3. PCR amplify Km^R fragment from pKD4 with the primers P1 and P2 through 30 cycles of 94 °C for 30 s, 58 °C for 60 s, 72 °C for 2 min followed by extension at 72 °C for 10 min using the PCR reaction mix shown in Table 2 (*see* **Note 3**).

Table 2
PCR reaction for individual fragments of Up, Km^R, and Dn

Template DNA	1.0 μl
Pfu DNA polymerase	1.0 μl
10× Pfu buffer	5.0 μl
dNTPs (2.5 mM each)	4.0 μl
Forward primer (20 μM)	1.0 μl
Reverse primer (20 μM)	1.0 μl
Distilled water	37.0 μl
Total volume	50.0 μl

4. Run a 1% agarose gel electrophoresis using 0.5× TAE buffer for the 3 DNA fragments.
5. Visualize the agarose gel under UV light to check the size of the fragments.
6. Cut out the gel pieces containing each of the three DNA fragments of the correct sizes.
7. Gel-purify each PCR product using QIAquick gel extraction kit following the manufacture's instruction.
8. Determine the concentration of gel-purified DNA fragments using Qubit fluorometer (Thermo Fisher). These purified DNA fragments are used as templates for the following overlapping extension PCR.

3.4 Overlapping Extension PCR to Join Three DNA Fragments

The goal of this step is to join all three fragments in the order of Up-Km^R-Dn fragments using overlapping PCR reaction. Although all three fragments can be joined at one overlapping extension PCR reaction, we found it is often unsuccessful and requires higher concentration of DNA fragments. Instead, we found that the strategy of joining the two adjacent fragments, which is then joined with the third fragment via two consecutive overlapping extension PCR reactions works efficiently and consistently in our experience. The following protocol is based on the two-step overlapping extension PCR reactions.

1. Join the two fragments, Km^R and Dn, using the PCR reaction mix in Table 3 with 30 cycles of 94 °C for 30 s, 58 °C for 60 s, 72 °C for 2 min followed by extension at 72 °C for 10 min (*see* **Note 4**).
2. Follow the **steps 4–8** in Subheading 3.3 for gel electrophoresis, gel-extraction, and measurement of DNA concentration of Km^R-Dn fragment.

Table 3
Overlapping extension PCR to join two fragments of Km^R and Dn

Template	
Gel-purified Km^R fragment	3.0 μl
Gel-purified Dn fragment	1.0 μl
Ex Taq polymerase	0.5 μl
10× Ex Taq buffer	5.0 μl
dNTPs (2.5 mM each)	4.0 μl
P1 primer (20 μM)	1.0 μl
Dn-R primer (20 μM)	1.0 μl
Distilled water	34.5 μl
Total volume	50.0 μl

Table 4
Overlapping extension PCR to join three fragments of Up and Km^R-Dn

Template	
Gel-purified Up fragment	4.0 μl
Gel-purified Km^R-Dn fragment	0.5 μl
Ex Taq polymerase	0.5 μl
10× Ex Taq buffer	5.0 μl
dNTPs (2.5 mM each)	4.0 μl
Up-F primer (20 μM)	1.0 μl
Dn-R priimer (20 μM)	1.0 μl
Distilled water	34.0 μl
Total volume	50.0 μl

3. Join Up fragment to Km^R-Dn fragment using the PCR reaction mix in Table 4 with 30 cycles of 94 °C for 30 s, 52 °C for 60 s, 72 °C for 2 min followed by extension at 72 °C for 10 min.
4. After completion of PCR cycle, repeat the **steps 4–8** in Subheading 3.3 for agarose gel electrophoresis and gel-extraction, and measurement of DNA concentration of the Up-Km^R-Dn fragment.

3.5 Electroporation with the Overlapping Extension PCR Products

The next step is to introduce the Up-Km^R-Dn fragment into *S.* Enteritidis cell that carry the plasmid pKD46 by electroporation. pKD46 has the genes for Red recombinase system under the control of arabinose-inducible promoter. When *Salmonella* carrying pKD46 is grown in the presence of L-arabinose, Red recombination system is expressed and mediates efficient homologous recombination of Up-Km^R-Dn fragment into the target site. Hence, the step begins with transformation of *S.* Enteritidis with pKD46.

3.5.1 Transformation of S. Enteritidis with pKD46

1. Streak an LB plate with *S.* Enteritidis strain.
2. Incubate at 37 °C overnight.
3. Lightly touch a single colony with a sterile loop and inoculate 10 ml LB medium. Incubate 37 °C overnight at 225 rpm.
4. Transfer 100 μl of the overnight culture to inoculate fresh 10 ml 2× YT medium.
5. Incubate at 37 °C, 250 rpm until OD_{600} reaches 0.8. It usually takes 3–5 h.
6. Centrifuge the culture at 3214 × *g*, 4 °C for 5 min. Carefully pour off the supernatant.
7. Wash pellet 3× with 100–200 μl of 10% cold glycerol.
8. Resuspend in 1 ml 10% cold glycerol.
9. Add 1 μl of plasmid pKD46 into the freshly prepared electrocompetent cells of *S.* Enteritidis and mix well by flicking gently.
10. Carefully transfer the cell/DNA mix into a chilled microelectroporation cuvette without introducing bubbles and ensure the cells are deposited across the bottom of the cuvette.
11. Electroporate the cells using ECM 399 Exponential Decay Wave Electroporation System and Voltage Booster/Gene Pulser Electroporator (Bio-Rad) using the parameters of 2.1 kV, 2450 kΩ, and 25 μF.
12. Immediately, add 500 μl of warm SOC medium (30 °C) into the cuvette. Gently mix by pipetting up and down and transfer into 2 ml round-bottomed culture tube.
13. Incubate at 250 rpm, 30 °C for 1 h (*see* **Note 5**).
14. Spread 100 μl of cells onto a prewarmed LB agar plate supplemented with Amp.
15. Incubate the plate at 30 °C overnight.

3.5.2 Transformation of S. Enteritidis Harboring pKD46 with Overlapping Extension PCR Product (Up- Km^R-Dn Fragment)

1. Prepare electrocompetent cells using *S.* Enteritidis strain transformed with pKD46 (from the **step 15** in Subheading 3.5.1) according to the **steps 1–5** in Subheading 3.5.1.
2. Induce the Red recombinase system by adding L-arabinose to the final concentration of 10 mM for either 1 h before harvesting or for the entire time of 3–5 h of culturing (*see* **Note 6**).
3. Electroporate the overlapping extension PCR product of three-joined fragment (Up-Km^R-Dn) into the competent cells *S.* Enteritidis (pKD46) as described in the **steps 9–13** in

Table 5
Colony PCR reaction

Template: single colony lysate	2.0 μl
GoTaq G2 green master mix	10.0 μl
BC-F (20 μM)	1.0 μl
BC-R (20 μM)	1.0 μl
Distilled water	6.0 μl
Total volume	20.0 μl

Subheading 3.5.1, except that (1) Up-Km^R-Dn fragment is used for transformation in place of pKD46, (2) LB agar plate supplemented with Km is used to recover transformants, and (3) incubation temperature is 37 °C (*see* **Note 7**).

3.6 Colony PCR to Check the Transformants

1. Lightly touch single colonies with sterile loops and suspend cells from each colony in 20 μl 1× PBS (*see* **Note 8**). Use the wild type strain as a negative control.
2. Boil at 95 °C for 10 min. Centrifuge at 3214 × *g* for 1 min and use the supernatant as the template for colony PCR.
3. Conduct a colony PCR for detection of the barcode insertion by using primer BC-F and BC-R, which anneal outside the Up-Km^R-Dn region (upstream of Up-F primer and downstream of Dn-R primer, respectively) using the PCR reaction mix in Table 5 through 30 cycles of 94 °C for 30 s, 58 °C for 2 min, 72 °C for 2 min followed by extension at 72 °C for 10 min (*see* **Note 9**).
4. After PCR, separate 10 μl of each PCR reaction on a 1% agarose gel.
5. Visualize the DNA fragments under UV light.
6. The PCR products from successfully transformed BC-tagged strains should be distinctive from the wild type used as a negative control (*see* **Note 10**).
7. Determine DNA sequences of the barcode in each positive strain by Sanger sequencing of the PCR product using appropriate primer flanking BC region (e.g., Up-F or Dn-R) as a sequencing primer.
8. Grow the barcode-tagged strains in LB medium (Km) at 37 °C overnight (*see* **Note 11**).
9. Make a glycerol stock and store at −80 °C for future use.

4 Notes

1. Other antibiotic resistance genes from different sources can replace Km^R gene in pKD4.
2. Appropriate insertion sites should be identified carefully for different bacterial species based on the available information on the genes and their functions in the bacterial species under the study.
3. We used Pfu DNA polymerase for this PCR step due to its high-fidelity and the blunt end created at the end of the resulting PCR products. Some thermostable DNA polymerases create A-overhangs at 3′ end of the products, which can interfere with efficient joining in overlapping extension PCR.
4. We found that TaKaRa Ex Taq polymerase performs best among multiple thermostable DNA polymerases we have evaluated for overlapping extension PCR.
5. Since pKD46 has temperature-sensitive replicon, the strains transformed with pKD46 should be incubated and kept $\leq$30 °C to maintain the plasmid.
6. Transformation efficiency is occasional very low, yielding only a few or no transformants. To increase the chance to get transformants, we prepared the electrocompetent cells in triplicate, which are then combined and used for each electroporation.
7. Transformation efficiency in this step is usually low, yielding no or only a few colonies. We usually keep the remaining transformation reaction at room temperature overnight. When there is no transformant colony found the next day, the remaining transformation reaction is plated on LB plates (Km), followed by overnight incubation at 37 °C. This procedure often yields positive transformant colonies.
8. We found the use of 1× PCR buffer instead of 1× PBS to suspend bacterial cells for colony PCR improves the efficiency of PCR. It is probably due to the prevention or reduction of depurination in DNA by 1× PCR buffer during boiling step [7]. For 1× PCR buffer, we use any extra 10× PCR buffer we can find in the freezer.
9. The use of primers annealing outside the Up-Km^R-Dn region in this step will not only confirm the presence of the Up-Km^R-Dn fragment in the chromosome, but also the insertion of the fragment in the correct chromosome locus.
10. The exact length of the PCR products from this step will depend on the design of the construction and primers, but it would be usually different between the wild type (negative

control) and the barcode-tagged strains, making it easy to identify the clones with successful BC-tagging.

11. If necessary, P22 transduction can be used to transfer the BC-tagged region into the fresh wild-type background.

References

1. Varble A, Albrecht RA, Backes S, Crumiller M, Bouvier NM, Sachs D et al (2014) Influenza a virus transmission bottlenecks are defined by infection route and recipient host. Cell Host Microbe 16:691–700. https://doi.org/10.1016/j.chom.2014.09.020
2. Lam LH, Monack DM (2014) Intraspecies competition for niches in the distal gut dictate transmission during persistent *Salmonella* infection. PLoS Pathog 10:e1004527. https://doi.org/10.1371/journal.ppat.1004527
3. Zhang T, Abel S, Abel Zur Wiesch P, Sasabe J, Davis BM, Higgins DE et al (2017) Deciphering the landscape of host barriers to *Listeria monocytogenes* infection. Proc Natl Acad Sci U S A 114:6334–6339. https://doi.org/10.1073/pnas.1702077114
4. Yang Y, Ricke SC, Tellz G, Kwon YM (2017) Quantitative tracking of *Salmonella* Enteritidis transmission routes using barcode-tagged isogenic strains in chickens: proof-of-concept study. Front Vet Sci 4:15. https://doi.org/10.3389/fvets.2017.00015
5. Yang Y, Tellez G, Latorre JD, Ray PM, Hernandez X, Hargis BM, Ricke SC, Kwon YM (2018) *Salmonella* excludes *Salmonella* in poultry: confirming an old paradigm using conventional and barcode-tagging approaches. Front Vet Sci 5:101. https://doi.org/10.3389/fvets.2018.00101
6. Datsenko KA, Wanner BL (2000) One-step inactivation of chromosomal genes in *Escherichia coli* K-12 using PCR products. Proc Natl Acad Sci U S A 97:6640–6645. https://doi.org/10.1073/pnas.120163297
7. Frost MR, Guggenheim JA (1999) Prevention of depurination during elution facilitates the reamplification of DNA from differential display gels. Nucleic Acids Res 27:e6

Chapter 14

Transposon-Aided Capture of Antibiotic Resistance Plasmids from Complex Samples

Sarah Delaney, Richard Murphy, and Fiona Walsh

Abstract

Transposon-aided capture of plasmids allows for antibiotic resistance plasmids to be acquired from complex samples. It is based on the insertion of a transposon, with a known origin of replication and selectable marker, into the plasmids present in a sample which can subsequently be captured after the removal of contaminating chromosomal DNA. Here we describe isolation of bacterial cells from a complex sample, DNA extraction, transposon-aided capture of plasmids in the sample, and analysis of the captured plasmids.

Key words Transposon, Plasmid, Antibiotic resistance, EZ-*Tn5*, Complex samples

1 Introduction

Plasmids are extrachromosomal DNA molecules capable of self-replication that exist naturally in bacterial cells [1]. They often carry genes that may benefit the survival of the organism, such as those conferring resistance to antibiotics [2]. Many plasmids also have the ability to conjugate to other bacteria, and are therefore considered major drivers in the spread of antibiotic resistance [3]. Plasmids comprise only a very small amount of the total DNA present in a complex sample. Also, only a fraction of the bacterial composition of a complex sample can be cultured in a laboratory environment [4]. Therefore, alternative methods of obtaining plasmids from these complex environments are required.

Firstly, bacterial cells are isolated from the complex sample. Then the DNA is extracted using an alkaline lysis method [5] to favor the isolation of plasmid DNA. Treatment of the extracted DNA with an exonuclease (plasmid-safe DNase) removes sheared chromosomal DNA and the remaining circular plasmid DNA undergoes a transposition reaction. The transposon-aided capture of plasmids method uses a known plasmid origin of replication (R6Kγori), which is functional in *Escherichia coli*, and selectable

Steven C. Ricke et al. (eds.), *Microbial Transposon Mutagenesis: Protocols and Applications*, Methods in Molecular Biology, vol. 2016, https://doi.org/10.1007/978-1-4939-9570-7_14,

marker (kanamycin resistance) on a *Tn5* derivative. Thus, after the reaction, the origin of replication is transposed onto the circular plasmids present [6]. The plasmids are captured by transformation to an *E. coli* host and subsequently selected on kanamycin, followed by plasmid isolation.

A major advantage of this method is that it has the capability to capture plasmids that do not have selectable markers and that may not have the ability to mobilize or replicate in a surrogate host species [7]. It also allows for plasmids from both Gram-positive and Gram-negative species to be captured and stably maintained in a surrogate host (*E. coli*) [8]. This method has been shown to favor the isolation of small plasmids [9], so it may give a misrepresentation of the total plasmid population present in a sample. This may be due to the decrease in transformation frequency as the size of the plasmid increases, meaning large plasmids will not transform as easily into the host *E. coli* [10]. Large plasmids are also more likely to be present in lower copy numbers compared to small plasmids making them more difficult to capture by this method. It has been suggested that plasmids which are unstable in *E. coli* or intractable by transposition are unlikely to be captured [8]. It is possible that linear plasmids would not be captured by this method, as they could be degraded by the exonuclease unless specialized enzymes are used and the *Tn5* origin of replication is not capable of replicating their extreme termini [11].

This method was originally developed for the isolation of plasmids from the human gut microbiome [12], and has also been used for studies into plasmids present in activated sludge [13], wastewater [14] and human dental plaque [6, 11]. These methods used a sequencing and primer walking approach to analyze the captured plasmids. In this manner, the transposon-aided capture of plasmids allowed for the identification of novel plasmids from these environments.

2 Materials

2.1 Isolation of Bacterial Cells

Tryptic soy broth (TSB): Weigh 15 g TSB and add 500 mL distilled water. Mix and autoclave at 121 °C for 15 min. Store at room temperature (*see* **Note 1**).

2.2 Alkaline Lysis

1. Resuspension buffer: 50 mM glucose, 25 mM Tris–HCl (pH 8.0), 10 mM EDTA (pH 8.0). Store at 4 °C, chill on ice before use.
2. Lysis buffer: 0.2 N NaOH, 1% (wt/vol) sodium dodecyl sulfate (SDS). Store at room temperature.
3. Neutralization buffer: 3 M potassium acetate, pH 4.8. Store at room temperature, prechill at 4 °C before use.

4. Isopropanol.
5. 70% ethanol.
6. MilliQ water.

2.3 Transposon-Aided Capture of Plasmids

1. Plasmid-Safe DNase (Epicentre).
2. MyTaq DNA polymerase and reaction buffer (Bioline).
3. 16S rRNA primers.
4. EZ-*Tn5* <R6Kγori/KAN-2> Insertion Kit (Epicentre).
5. Sterile water.
6. TransforMax EC100D pir-116 Electrocompetent *Escherichia coli* (Epicentre).
7. Luria–Bertani (LB) Agar: Weigh 10 g of LB broth (Duchefa-Biochemie) and 5 g of Micro Agar (Duchefa-Biochemie). Add 500 mL of distilled water and mix. Autoclave at 121 °C for 15 min. Store at 55 °C.
8. Kanamycin (Sigma).

2.4 Analysis of Captured Plasmids

1. Luria–Bertani (LB) Broth: Weigh 10 g of LB broth (Duchefa-Biochemie), add 500 mL of distilled water and mix. Autoclave at 121 °C for 15 min. Store at room temperature.
2. Qiagen HiSpeed Plasmid Midi kit.
3. 1% agarose gel.
4. GelRed (Biotium).
5. QIAquick Gel Extraction Kit (Qiagen).
6. Electrocompetent *Escherichia coli* cells.
7. Luria–Bertani (LB) Agar: Weigh 10 g of LB broth (Duchefa-Biochemie) and 5 g of Micro Agar (Duchefa-Biochemie). Add 500 mL of distilled water and mix. Autoclave at 121 °C for 15 min. Store at 55 °C.
8. Macherey-Nagel NucleoSpin Plasmid kit.
9. *Eco*RI restriction enzyme (Thermo Scientific).
10. Antibiotic disks (Oxoid).

3 Methods

3.1 Isolation of Bacterial Cells

1. Separate bacterial cells from the complex samples (*see* **Note 2**) by adding 0.1 g of the complex sample to 0.9 mL of nonselective TSB and incubate at room temperature on a rocker at 70 rpm overnight (*see* **Note 3**).
2. Allow to settle and collect the supernatant containing the bacterial fraction (around 0.8 mL) (*see* **Note 4**). Centrifuge at 2800 × *g* for 10 min. Discard the supernatant.

3.2 Alkaline Lysis

1. Resuspend the bacterial pellet in 100 μL ice-cold resuspension buffer.
2. Lyse bacterial cells with 200 μL of lysis buffer for 4 min. Add 150 μL of neutralization buffer. Centrifuge the samples at 18,300 × *g* for 10 min at 4 °C.
3. Mix the supernatant containing plasmid DNA with an equal volume of isopropanol and incubate at −20 °C for 15 min. Centrifuge samples at 18,300 × *g* for 30 min at 25 °C. Remove the supernatant, add 500 μL of 70% ethanol to the pellet and centrifuge at 18,300 × *g* for 5 min at 25 °C. Resuspend the pellet in 50 μL MilliQ water.

3.3 Transposon-Aided Capture of Plasmids

1. Remove sheared chromosomal DNA by treating with Plasmid-Safe DNase (Epicentre), according to the manufacturer's guidelines.
2. Check for the presence of chromosomal DNA by testing for the presence of the 16S rRNA genes by PCR (*see* **Note 5**).
3. Insert the transposon onto the plasmids using the EZ-*Tn5* <R6Kγori/KAN-2> Insertion Kit (Epicentre), according to the manufacturer's guidelines (*see* **Note 6**).
4. Dilute the 50 μL reaction with 450 μL sterile water and purify with Vivaspin 500 MWCO 100,000 Protein Concentrator Spin Columns (GE Healthcare Life Sciences), which reduces the reaction volume to 10 μL.
5. Electroporate 5 μL of the reaction mixture into 100 μL TransforMax EC100D pir-116 Electrocompetent *Escherichia coli* (Epicentre) (*see* **Note 7**) at 1.8 kV. Spread the transformed cells (100 μL) onto LB agar plates supplemented with 50 mg/L kanamycin to select for EZ-*Tn5*. Incubate overnight at 37 °C (*see* **Note 8**).

3.4 Analysis of Captured Plasmids

1. Inoculate single colonies from the agar plates into 150 mL LB broth. Incubate overnight at 37 °C with shaking at 225 rpm.
2. Extract plasmid DNA using the Qiagen HiSpeed Plasmid Midi kit according to the manufacturer's guidelines (*see* **Note 9**). Visualize plasmids by loading 10 μL onto a 1% agarose gel stained with GelRed (Biotium) and run at 70 V for 60 min.
3. Extract and purify bands of plasmid DNA from the agarose gel using the QIAquick Gel Extraction Kit (Qiagen) according to the manufacturer's guidelines (*see* **Note 10**) (Fig. 1). Electroporate the individual bands of plasmid DNA (5 μL) into *Escherichia coli* DH5α (25 μL) (or similar lab strain) at 1.8 kV.
4. Spread the transformed cells onto LB agar plates supplemented with selected antibiotics (*see* **Note 11**). Incubate at 37 °C overnight.

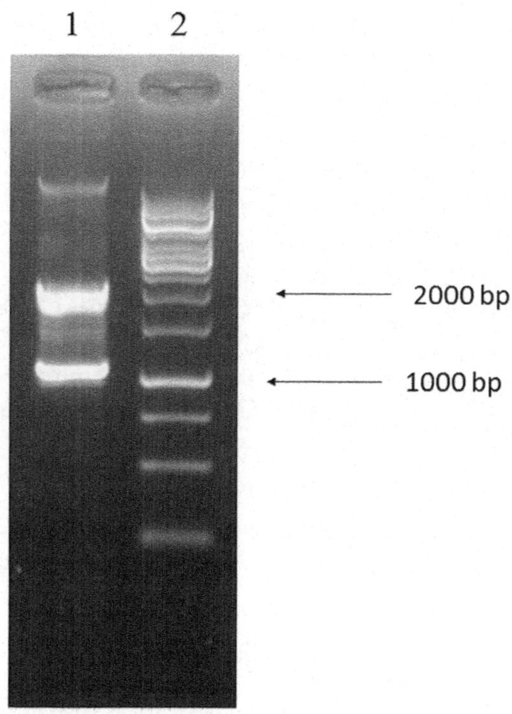

Fig. 1 (1) Plasmid DNA extracted from transformants after the transposition reaction. The two brightest bands were extracted and analyzed further. (2) 1 kb ladder

5. Inoculate single colonies from the agar plates into 5 mL LB broth. Incubate overnight at 37 °C with shaking at 225 rpm. Extract plasmid DNA using the Macherey-Nagel NucleoSpin Plasmid kit according to the manufacturer's guidelines. Visualize plasmids on a 1% agarose gel stained with GelRed.
6. Digest plasmid DNA with *Eco*RI restriction enzyme (Thermo Scientific) according to the manufacturer's guidelines and perform antibiotic susceptibility testing on the transformants via the disk diffusion method according to the Clinical Laboratory Standards Institute guidelines [15] to determine the resistance profile and similarity of the plasmids obtained.

4 Notes

1. TSB is a nonselective media and will allow for the enrichment of bacteria in the complex sample.
2. Alternatively, to avoid the enrichment step, add 0.9 mL of 0.85% NaCl to the complex sample instead of TSB. Briefly vortex to mix, allow to settle and collect the supernatant containing the bacterial fraction.

3. We used cecal samples from broiler chickens as a representative of a complex sample. This method can be performed with other complex samples in the same manner.
4. If the supernatant is too thick and does not form a separate layer from the sample increase the volume of broth.
5. We performed the 16S rRNA PCR using the following primers: Forward 5′-TCGTCGGCAGCGTCAGATGTGTATAAGAGACAGCCTACGGGNGGCWGCAG-3′ and Reverse 5′-GTCTCGTGGGCTCGGAGATGTGTATAAGAGACAGGACTACHVGGGTATCTAATCC-3′; [16] and under the following conditions: 95 °C for 3 min; 35 cycles of: 95 °C for 30 s, 55 °C for 30 s, 72 °C for 30 s; and finally 72 °C for 5 min.
6. This is a 2 h reaction where the transposon is randomly inserted into the plasmid DNA.
7. The *E. coli* used for transformation must be *rec*A negative and express the *pir* gene product.
8. At this point, you can proceed with sequencing and primer walking as an alternate way of analyzing of the captured plasmids.
9. The concentration of plasmid DNA extracted using a mini-kit was too low for gel extraction therefore a midi-kit was used to obtain a higher concentration.
10. We extracted two bands from the agarose gel; band 1 and band 2.
11. We used ampicillin (32 mg/L), tetracycline (16 mg/L), kanamycin (25 mg/L), colistin (16 mg/L), and ciprofloxacin (4 mg/L). Transformants were obtained on ciprofloxacin with DNA from band 1 and on ampicillin, tetracycline, and ciprofloxacin with DNA from band 2.

References

1. Carattoli A (2013) Plasmids and the spread of resistance. Int J Med Microbiol 303:298–304
2. Bennett PM (2008) Plasmid encoded antibiotic resistance: acquisition and transfer of antibiotic resistance genes in bacteria. Br J Pharmacol 153:S347–S357
3. Munita JM, Arias CA (2016) Mechanisms of antibiotic resistance. Microbiol Spectr 4(2)
4. Kav AB, Benhar I, Mizrahi I (2013) A method for purifying high quality and high yield plasmid DNA for metagenomics and deep sequencing approaches. J Microbiol Methods 95:272–279
5. Birnboim HC, Doly J (1979) A rapid alkaline extraction procedure for screening recombinant plasmid DNA. Nucleic Acids Res 7 (6):1513–1523
6. Mullany P, Allan E, Warburton PJ (2012) Tetracycline resistance genes and mobile genetic elements from the oral metagenome. Clin Microbiol Infect 18(Suppl. 4):58–61
7. Delaney S, Murphy R, Walsh F (2018) A comparison of methods for the extraction of plasmids capable of conferring antibiotic resistance in a human pathogen from complex broiler cecal samples. Front Microbiol 9:1731. https://doi.org/10.3389/fmicb.2018.01731
8. Dib JR, Wagenknecht M, Farías ME, Meinhardt F (2015) Strategies and approaches in plasmidome studies—uncovering plasmid diversity disregarding of linear elements? Front Microbiol 6:463. https://doi.org/10.3389/fmicb.2015.00463

9. Jørgensen TS, Kiil AS, Hansen MA, Sørensen SJ, Hansen LH (2015) Current strategies for mobilome research. Front Microbiol 5:750. https://doi.org/10.3389/fmicb.2014.00750
10. Szostková M, Horáková D (1998) The effect of plasmid DNA sizes and other factors on electrotransformation of *Escherichia coli* JM109. Bioelectrochem Bioenerg 47:319–323
11. Warburton PJ, Allan E, Hunter S, Ward J, Booth V, Wade WG, Mullany P (2011) Isolation of bacterial extrachromosomal DNA from human dental plaque associated with periodontal disease, using transposon-aided capture (TRACA). FEMS Microbiol Ecol 78 (2):349–354
12. Jones BV, Marchesi JR (2007) Transposon-aided capture (TRACA) of plasmids resident in the human gut mobile metagenome. Nat Methods 4:51–61
13. Zhang T, Zhang XX, Ye L (2011) Plasmid metagenome reveals high levels of antibiotic resistance genes and mobile genetic elements in activated sludge. PLoS One 6(10):e26041. https://doi.org/10.1371/journal.0026041
14. Burmølle M, Norman A, Sørensen SJ, Hestbjerg Hansen L (2012) Sequencing of IncX-plasmids suggests ubiquity of mobile forms of a biofilm-promoting gene cassette recruited from *Klebsiella pneumoniae*. PLoS One 7(7): e41259. https://doi.org/10.1371/journal.pone.0041259
15. Clinical and Laboratory Standards Institute (2016) Performance Standards for Antimicrobial Susceptibility. CLSI Document M100S. Clinical and Laboratory Standards Institute, Wayne, PA
16. Klindworth A, Pruesse E, Schweer T, Peplies J, Quast C, Horn M, Glöckner FO (2013) Evaluation of general 16s ribosomal RNA gene PCR primers for classical and next-generation sequencing-based diversity studies. Nucleic Acids Research 41(1):e1–e1

Chapter 15

Efficient Gene Deletion Method for *Listeria monocytogenes*

Hossam Abdelhamed, Attila Karsi, and Mark L. Lawrence

Abstract

Inactivation or deletion of genes allows for investigation and understanding of gene function. To facilitate markerless gene deletion in *Listeria monocytogenes*, we developed a new suicide plasmid (pHoss1). pHoss1 contains the pMAD backbone, the *secY* antisense cassette from pIMAY driven by an inducible Pxyl/tetO promoter, a heat-sensitive origin of replication, four unique restriction sites (*SalI*, *EcoRI*, *SmaI*, and *NcoI*), and erythromycin resistance gene. We demonstrated that pHoss1 is very efficient for introducing mutations into different *L. monocytogenes* strains. In this chapter, we include a brief description of pHoss1 and the method used for gene deletion in *L. monocytogenes* using pHoss1.

Key words *Listeria monocytogenes*, pHoss1, Pxyl/tetO promoter, *secY*, Gene deletion

1 Introduction

Generating targeted gene deletions and replacements is a fundamental approach for the characterization of gene function and production of vaccine candidates [1, 2]. Conventional allelic exchange relies on the availability of selectable marker(s) on the plasmid allowing positive selection of transformants and subsequent chromosomal integration. The first allelic exchange results in a merodiploid with a copy of the gene deletion and the intact gene. A second allelic exchange is required to resolve the merodiploid and result in a gene deletion strain. To identify gene deleted bacteria, screening for loss of plasmid selectable marker is required unless the plasmid carries a second marker that enables positive selection for plasmid loss [3].

A number of plasmids have been developed for introducing mutations in Gram-positive bacteria such as pAUL-A, pLSV2, pMAD, pKOR1, and pIMAY [4–8]. In our hands, these plasmids were not efficient for introducing markerless gene deletions in *L. monocytogenes*. Some require screening for loss of plasmid. Others carry a second selectable marker for plasmid loss, but they have low transformation efficiency in *L. monocytogenes*. Thus, we

Steven C. Ricke et al. (eds.), *Microbial Transposon Mutagenesis: Protocols and Applications*, Methods in Molecular Biology, vol. 2016, https://doi.org/10.1007/978-1-4939-9570-7_15,

developed a new suicide plasmid, pHoss1, with positive selection for plasmid loss and efficient generation of gene deleted strains [9].

pMAD was constructed for generating allelic replacements in *Staphylococcus aureus*, *Listeria monocytogenes*, and *Bacillus cereus* [6]. This vector has a temperature sensitive origin of replication, an erythromycin selection marker, and a *lacZ* gene encoding β-galactosidase (*bgaB*) for blue–white screening of the second allelic exchange. When we used pMAD, color development at room temperature was slow (10 days) in *L. monocytogenes*, and color selection sometimes resulted in false positives. pKOR1 and pIMAY were developed for allelic exchange in *Staphylococcus* [7, 10]. These two vectors contain antisense *secY* RNA expression for positive selection for the second allelic exchange and plasmid loss. Although pIMAY and pKOR1 were used successfully for allelic replacement in *S. aureus*, our transformation efficiencies of these two plasmids were very low in *L. monocytogenes*.

To construct pHoss1, we adapted pMAD by replacing the *bga*B fragment with an antisense *secY* RNA expression cassette under the control of an inducible Pxyl/tetO promoter (Fig. 1). As previously described [9], the antisense *secY* RNA expression cassette was

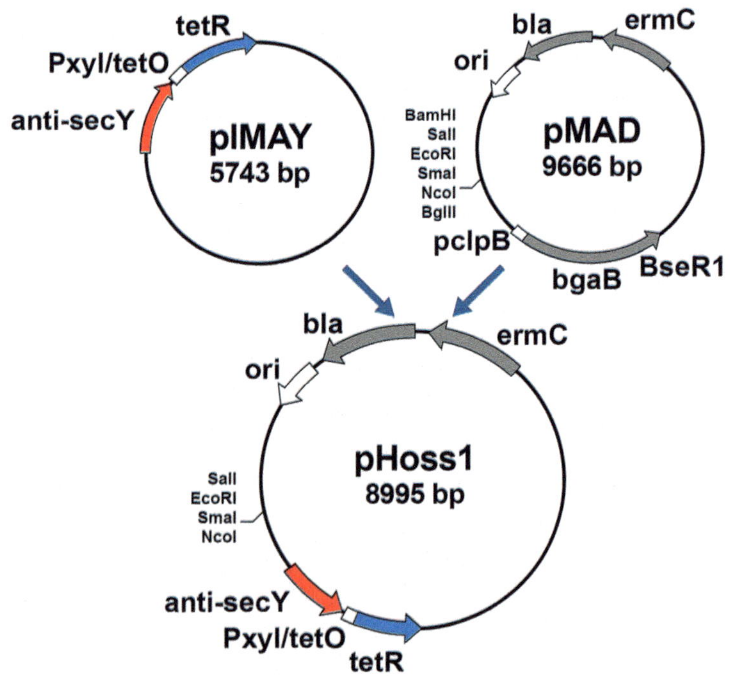

Fig. 1 Construction of pHoss1 suicide plasmid [9]. Tetracycline-inducible antisense *secY* region (*anti-secY*) was amplified from pIMAY and ligated with pMAD plasmid backbone. Abbreviations: *MCS* Multiple cloning site, *ermC* Erythromycin resistance gene, *bla* Beta-lactamase gene), *ori* Replication origin of the plasmid from pBR322

obtained by PCR amplification from pIMAY with the Anti-BseRI-F01 (AAGAGGAGGATCTAATGAT TCAAACCCTTGTG) and Anti-BglII-R01 (AAAGATCTTGAAGTTACCATCACGGAAAAA GG) primers. The amplified antisense *secY* RNA fragment was digested with *BseR*I and *Bgl*II and ligated into pMAD digested with the same two enzymes (7624 bp fragment of pMAD). We found that expression of the *secY* antisense RNA eliminates growth of *L. monocytogenes* with chromosomally integrated pHoss1. Thus, induction with anhydrotetracyline provided strong selection for the second allelic exchange and plasmid loss. Approximately 80% of the resulting colonies were desired deletion mutants (the remaining 20% reverted to wild-type genotype).

2 Materials

2.1 Bacterial Strains and Plasmid

1. *Escherichia coli* strains DH5α or TOP10 for cloning and maintenance of pHoss1.
2. pHoss1 (available at https://www.addgene.org/).

2.2 Growth Media

1. LB agar (Difco, Sparks, Maryland) to grow *E. coli* strains.
2. LB broth (Difco) to grow *E. coli* strains.
3. Brain Heart Infusion Agar (Difco) to grow *Listeria* strains.
4. Brain Heart Infusion broth (Difco) to grow *Listeria* strains.
5. SOB medium (Sigma-Aldrich, St. Louis, MO) for use in transformation.

2.3 Reagents, Enzymes, and Kits

1. Taq DNA polymerase (Promega, Madison, WI).
2. Appropriate restriction endonucleases (New England Biolabs, Ipswich, MA). *Sal*I, *Eco*RI, *Sma*I, and *Nco*I sites are available in pHoss1.
3. Sucrose (Sigma-Aldrich). Prepare 500 mM sucrose solution by dissolving 17.1 g sucrose in 100 mL water. Sterilize by filtration through 0.2 μm filter.
4. 1 mM HEPES.
5. TE buffer, pH 8.0, autoclaved.
6. Milli-Q Ultrapure water (or equivalent deionized water).
7. 100% (200 proof) ethanol.
8. 1× TAE buffer.
9. Agarose, molecular biology grade (Sigma-Aldrich).
10. DNA ladder (1 kb Plus DNA Ladder, Invitrogen).
11. DNA gel-extraction kit (Qiagen).
12. Electroporation cuvette (0.2 cm).

13. ExoSAP-IT (Thermo Fisher Scientific) for cleanup of amplified PCR product.
14. Wizard Genomic DNA Kit (Promega) for isolation of genomic DNA from *Listeria* strains.
15. QIAprep Spin Miniprep Kit (Qiagen) for plasmid isolation from *E. coli* cultures.
16. QIAquick gel extraction Kit (Qiagen) for purification of DNA bands produced by PCR or restriction endonuclease digestion after separation on an agarose gel.

2.4 Antibiotics

1. Ampicillin (100 μg/mL) to select and maintain pHoss1 in *E. coli*.
2. Erythromycin (10 μg/mL) to select for *L. monocytogenes* with chromosomally integrated pHoss1 (*see* **Note 1**).
3. Anhydrotetracycline (ATc, 1.5 μg/mL final concentration) for induction of *secY* antisense RNA to provide positive selection for second allelic exchange and loss of pHoss1 in *L. monocytogenes*.

3 Methods

The methods described below for constructing gene deletions in *L. monocytogenes* are also summarized in Fig. 2.

3.1 Designing Primers for the *L. monocytogenes* Target Gene

1. Design four PCR primers (A, B, C, and D). Primer3 is effective for primer design (http://frodo.wi.mit.edu) (*see* **Note 2**).
2. Add restriction endonuclease sequences to the two flanking primers (A and D) for cloning. pHoss1 contains four cloning sites (*Sal*I, *Eco*RI, *Sma*I, and *Nco*I). If a different restriction endonuclease is used in each flanking primer, dephosphorylation of linear vector will not be needed, and vector self-ligation problem will be avoided, yielding increased ligation efficiency.
3. Some number of nucleotides between restriction sites and 5′ end of the primer (preceding the restriction endonuclease sequences) are required to increase digestion efficiency. Ensure that the bases do not result in the formation of a hairpin structure (*see* **Note 3**).
4. Add the reverse complement of primer B to the 5′ end of primer C to create an overlap region that enables fusion of AB fragment to CD fragment during the second round of PCR [11].

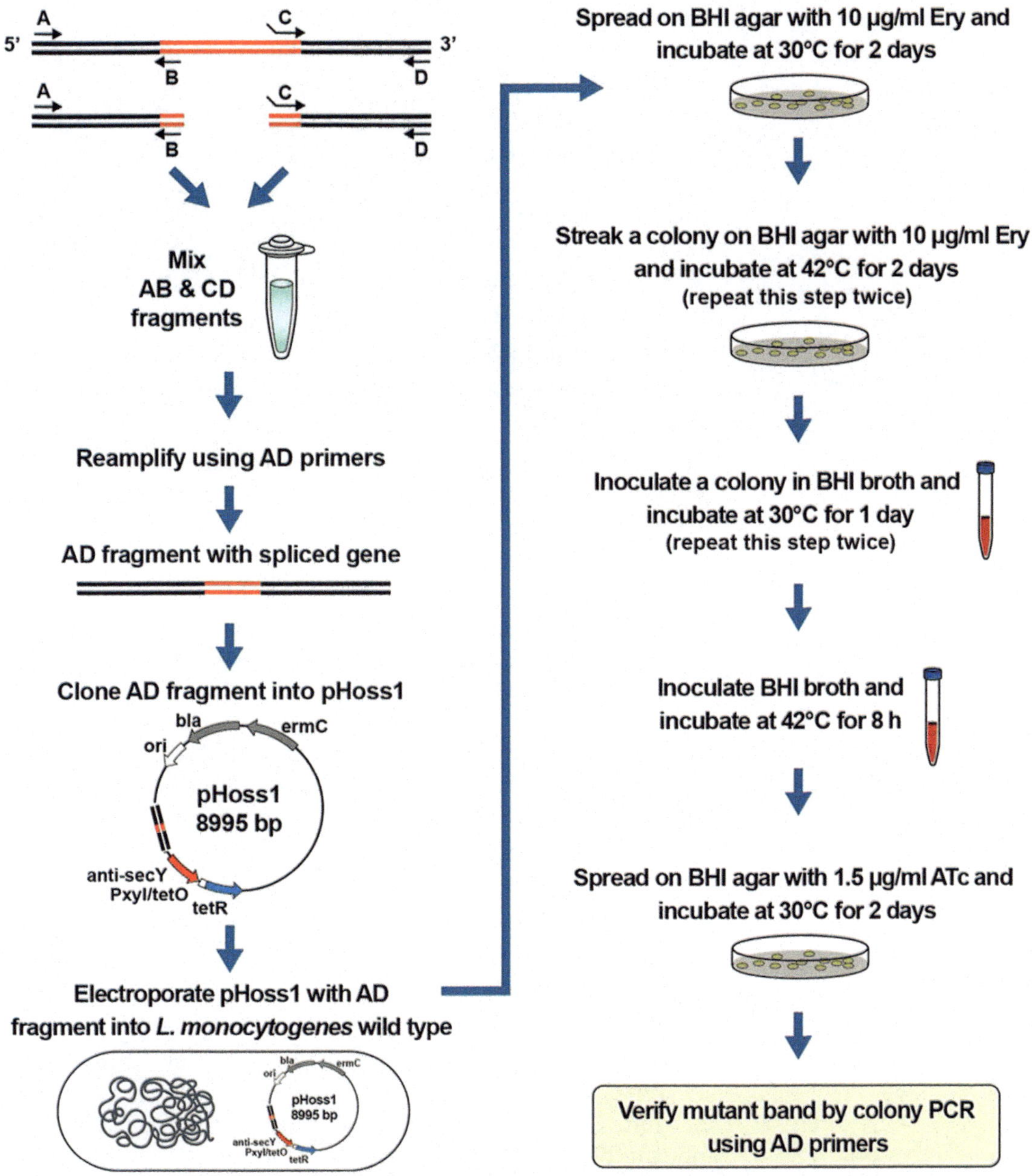

Fig. 2 Strategy for gene deletion in *Listeria monocytogenes* using pHoss1. Following transformation into *L. monocytogenes*, transformants with plasmid are selected at the permissive temperature (30 °C), then chromosomal integration of the plasmid is selected at the nonpermissive temperature (42 °C). Positive selection for plasmid excision and loss is conducted at 30 °C in the presence of anhydrotetracycline (ATc), leaving the gene deletion on the chromosome. In the final step, deletion mutations are verified by PCR using primers A and D

3.2 Construction of *L. monocytogenes* Gene Deletion Fragments

1. Amplify approximately 1 kb fragments from the upstream and downstream of the target gene separately with (A/B) and (C/D) primers (*see* **Note 4**).
2. PCR reaction (50 μL) include 50–100 ng *L. monocytogenes* genomic DNA as template, 1.25 U Taq DNA polymerase, 1.5 mM $MgCl_2$, 0.2 mM primers, and 0.2 mM dNTP mix. The thermocycler conditions include an initial denaturation at 94 °C for 5 min, 30 cycles of 94 °C for 30 s, 55 °C for 1 min, 72 °C for 1 min, and a final extension at 72 °C for 10 min.
3. Run 1/10 of the PCR reaction on a 1% agarose gel to confirm the presence and size of amplified fragment.
4. Mix equal volumes of upstream and downstream PCR products and dilute 1:20.
5. Generate an overlap extension fragment with A and D primers using diluted AB and CD fragment mixture as template. Thermocycler conditions include: an initial denaturation at 94 °C for 5 min, 30 cycles of 94 °C for 30 s, 55 °C for 2 min, 72 °C for 2 min, and a final extension at 72 °C for 10 min.
6. Purify PCR product by separating the DNA using agarose gel electrophoresis and excise the band using QIAquick Gel Extraction Kit (*see* **Note 5**).

3.3 Cloning Gene Deletion Fragment into pHoss1

1. Digest both the vector and overlap extension fragment (insert) with the appropriate restriction endonucleases. For efficient cutting, incubate digestion mix at the appropriate temperature (usually 37 °C except for *Sma*I at 25 °C) for 4–8 h (*see* **Note 6**).
2. Purify the vector and insert by separating the DNA using agarose gel electrophoresis and excising the appropriate bands using QIAquick Gel Extraction Kit.
3. Analyze agarose gels under long-wave UV. Minimize exposure of fragments to UV light that may cause DNA damage.
4. Set up the ligation reaction (10 μL) using a molar ratio of 1:4 linearized pHoss1 to insert with T4 DNA ligase in 1× buffer solution. Gently mix the reaction by pipetting up and down and spin briefly (*see* **Note** 7).
5. Incubate the ligation reaction at 16 °C overnight.
6. Chill on ice and transform 1–5 μL of the reaction into 50 μL competent *E. coli* DH5α.

3.4 Transformation of pHoss1 with Gene Deletion Fragment into *E. coli* DH5α

1. Thaw 50 μL competent *E. coli* DH5α on ice for each transformation (*see* **Note 8**).
2. Add 1–5 μL ligation reaction mixture to the competent bacteria and swirl to mix the DNA and bacteria. Place the tube on ice for 20–30 min.

3. Apply heat shock by placing the tube at 42 °C for 60–90 s.
4. Immediately transfer the tube on ice for 2 min.
5. Add 400 μL of SOC medium to the tubes and incubate at 37 °C, with shaking, for 1 h.
6. Spread 200 μL of culture onto prewarmed LB agar plates with ampicillin and incubate overnight at 37 °C (*see* **Note 9**).

3.5 Verification of the Insert by Colony PCR and Sequencing

1. Colony PCR is effective for screening transformant colonies to identify clones with integrated plasmid (*see* **Note 10**). Screen 20 colonies for identification of positive clones. Usually 20 colonies are sufficient to identify a correct insert.
2. Pick a single colony from LB agar plates with a sterile pipette tip and suspend in 20 μL of sterile water. Boil for 10 min at 100 °C to lyse the bacteria and release plasmid DNA.
3. Prepare PCR master mix solution for 20 reactions in total volume of 25 μL per reaction. The master mix includes all the components required for PCR except DNA template.
4. Run the following PCR cycling conditions: an initial denaturation at 94 °C for 10 min, 30 cycles of 94 °C for 30 s, 55 °C for 2 min, 72 °C for 2 min, and a final extension at 72 °C for 10 min.
5. Separate PCR fragments using 1% agarose gel electrophoresis to identify clones containing the desired insert.
6. Save the pipette tip or leftover bacteria–water suspension to start cultures of positive clones.
7. After identifying positive clones, inoculate 5 mL LB broth plus ampicillin and incubate overnight at 37 °C.
8. On the next day, purify plasmid from 2 mL overnight culture of positive clones by QIAprep Spin Miniprep Kit. The final volume of miniprep DNA should be 50 μL in sterile water or elution buffer. Prepare frozen stock from the remaining overnight cultures.
9. Submit 3–4 positive plasmids for sequencing to confirm insert orientation and the junction sequences between the plasmid and insert DNA (*see* **Note 11**).

3.6 Preparation of Competent L. monocytogenes

1. Grow 5 mL overnight culture of *L. monocytogenes* in BHI broth at 37 °C in a shaking incubator.
2. On the next day, inoculate 3 mL of overnight culture into 300 mL BHI/0.5 M sucrose in a 1 L flask. Grow in a shaking incubator at 220 rpm at 37 °C until OD_{600} reaches 0.4–0.5.
3. Incubate the culture for 15–20 min on ice to stop bacteria growth. Bacteria should be kept cold for all remaining steps.

4. Harvest bacteria by centrifugation at 4000 × *g* for 10 min at 4 °C and discard the supernatant.
5. Resuspend bacterial pellet in 100 mL HEPES/Sucrose. Do this by adding 20 mL first and swirling bacteria on ice. Once resuspended, add remaining 80 mL. Harvest again by centrifugation at 4000 × *g* for 10 min, 4 °C, and decant supernatant. Repeat this step two more times.
6. Resuspend bacteria in HEPES/Sucrose by inverting several times and incubate bacteria on ice for 1–2 h to facilitate genetic transformation. Harvest by centrifugation and decant supernatant.
7. Resuspend pellet in 3 mL of ice-cold HEPES/Sucrose containing 10% glycerol.
8. Transfer 50 μL aliquots of bacterial suspension to freezer tubes and store them at −80 °C for future use.

3.7 Electroporation of pHoss1 into *L. monocytogenes*

1. Remove electroporation competent *L. monocytogenes* from −80 °C freezer and thaw carefully on ice.
2. Add 5 μL (0.1–50 ng of DNA) of recombinant pHoss1 vector to 50 μL of competent bacteria. Mix by tapping the tube gently and incubate on ice for 20–30 min.
3. Transfer the bacteria–DNA suspension into a chilled electroporation cuvette (0.2 cm^2) and tap the cuvette gently to settle the mixture to the bottom.
4. Set the electroporation apparatus to the following settings by referring to the instructions provided with the instrument: 200 Ω, 2.5 kV, 25 μF.
5. Place cuvette into the electroporation chamber until the cuvette connects with the electrical contacts.
6. Pulse the sample once, then quickly remove the cuvette. Immediately add 0.5 mL of BHI broth with 500 mM sucrose and pipet up and down to cool the bacteria.
7. Transfer the bacterial suspension to a sterile 5-mL culture tube. Recover transformed bacteria by incubation for 3 h at 30 °C while shaking at 225–250 rpm (*see* **Note 12**).
8. After 3 h, spread 250 μL of bacterial suspension on BHI agar with erythromycin (10 μg/mL) and incubate plates at 30 °C for 2 days. The permissive temperature of 30 °C allows the plasmid to replicate in the bacteria (*see* **Note 13**).
9. Pick 10 colonies from BHI plates and inoculate individual 5 mL BHI broth plus erythromycin cultures. Grow them at 30 °C overnight with shaking at 200 rpm (*see* **Note 14**).

10. On the next day, perform colony PCR analysis of the erythromycin-resistant colonies using A and D primers to verify transformants are carrying plasmid (*see* **Note 15**).

3.8 Integration of pHoss1 and Selection of Mutants

1. To identify bacteria with chromosomally integrated plasmid, streak a single colony of *L. monocytogenes* with recombinant plasmid on a BHI agar plate with erythromycin and incubate at 42 °C for 2 days. This procedure should be repeated three times at 42 °C. Each time, one colony is streaked to a new plate. The nonpermissive temperature (42 °C) inhibits plasmid replication; therefore, only colonies with chromosomal plasmid integration can grow in the presence of erythromycin (*see* **Note 16**).
2. Confirm integration of plasmid in *L. monocytogenes* chromosomal DNA by colony PCR. The meroploid intermediates contain both wild-type and deleted alleles.
3. Next, inoculate two or three colonies from BHI erythromycin plates into 1–2 mL BHI broth without antibiotics (no erythromycin) and grow overnight at 30 °C in shaking incubator at 200 rpm. Repeat this step twice by inoculating 1–2 mL of fresh BHI broth with 20 μL of overnight culture.
4. After overnight incubation of the third culture, increase the incubation temperature to 42 °C for an additional 8 h.
5. To select deletion mutants with excised plasmid by a second allelic exchange, spread 50 μL of several serial dilutions from the culture on a BHI agar plate containing 1.5 μg/mL ATc. Incubate plates at 30 °C for 3 days (*see* **Note 17**).
6. Pick 20 colonies from BHI plate with ATc and perform colony PCR using primers A and D. Include PCR amplification from genomic DNA of wild-type as a control (*see* **Note 18**). Deletion mutants will amplify a single PCR fragment that is smaller than the wild type PCR fragment.
7. Confirm deletion mutants are erythromycin sensitive by inoculating into BHI broth with erythromycin and incubating overnight.

4 Notes

1. Because ampicillin is a drug of choice used for treatment of a listeriosis infection [12], it is undesirable to use ampicillin to select transformation of plasmid into *Listeria monocytogenes.* We recommend erythromycin to select transformation of plasmid into *L. monocytogenes.*
2. To generate complete gene deletion, the sequences from start to stop codons should be removed. Therefore, B and C primers

should be located immediately upstream and downstream of stop and start codons, respectively. In some cases when a suitable primer cannot be designed, B and C primers can be designed 1–50 bases after the start and before the stop codon, respectively.

3. New England Biolabs Inc. (NEB) determined the optimum number of nucleotides required for each restriction endonuclease. The table can be found through this link: https://www.neb.com/~/media/NebUs/Files/Chart%20image/cleavage_olignucleotides_old.pdf
4. The allelic exchange procedure worked well when amplified sequences flanking the deletion were about 750–1000 bp in length.
5. When running gel electrophoresis for purification purposes, it is important to have strong bands and to have separation between bands. This can be achieved by using a wide gel comb, running the gel on lower voltage over longer time, and skipping lanes between samples. In addition to a DNA ladder standard, we recommend including an uncut control pHoss1 vector as a control.
6. When digesting pHoss1 with a single restriction endonuclease, it is important to treat the digested pHoss1 with a phosphatase prior to the gel purification step to prevent recircularization of the vector. Because DNA is lost during the gel purification step, it is critical to digest enough insert and plasmid. We recommend using the entire PCR reaction (20 μL) and plasmid preparation (50 μL). It is also important that as much of the plasmid as possible be cut with both enzymes. To achieve a completely cut sample, it is important that the digest is incubated for at least 4 h or overnight. We recommend High-Fidelity restriction endonucleases to increase digestion efficiency and to simplify identification of compatible buffer.
7. It is difficult to calculate the plasmid and insert ratio based on DNA concentration alone. We recommend conducting multiple ligations with varying plasmid and insert ratios.
8. Preparation of *E. coli* competent bacteria could be performed by $CaCl_2$ treatment and transformation as described previously [13].
9. After overnight incubation at 37 °C, it is important to take the plates out from the incubator and store at 4 °C. If the plates are left in the incubator too long, small satellite colonies (bacteria with no plasmid) will form around larger transformed colonies, which may make picking positive colonies difficult.
10. Plasmid with insert and plasmid without insert will both confer ampicillin resistance to the transformed *E. coli* DH5α. Colony

PCR can discern between bacteria carrying plasmid with insert and plasmid without insert.

11. We recommend cleanup of the amplified PCR product with ExoSAP-IT before sequencing.
12. Allow the transformed bacteria to recover prior to spreading on BHI agar with erythromycin. A minimum of 3 h recovery allows time for expression of erythromycin resistance.
13. The entire transformation can be divided and spread on two agar plates. Transformants will generally be visible in 3–4 days at 30 °C, and 5–20 colonies are expected. The earliest that colonies will be visible is usually 2 days. If the colonies are small, allow them to grow an additional day at 30 °C.
14. Transformation efficiency of pHoss1 into *L. monocytogenes* is high. All erythromycin-resistant colonies after electroporation should be PCR positive with primers A and D.
15. PCR should result in two bands confirming wild-type (high molecular weight band) and mutated (lower molecular weight band) alleles.
16. pHoss1 is a temperature-sensitive vector. After transformation of pHoss1 into *L. monocytogenes*, integration of the plasmid into the chromosome is selected during growth at the nonpermissive temperature (42–44 °C) under erythromycin selective pressure. Subsequent growth of the cointegrates at the permissive temperature (30 °C) enables a second allelic exchange event.
17. Positive selection for plasmid loss is based on antisense *secY* RNA expression, which is effective in *L. monocytogenes* [9]. *L. monocytogenes* is sensitive to more than 2 μg/mL ATc. Thus, we used 1.5 μg/mL ATc for induction of the *secY* antisense RNA.
18. The expected size of the amplified fragment in the mutant strain is about 2000 bp (depending on the size of the amplified flanking regions), while the size in the wild-type strain is about 2000 bp plus the size of the deleted gene.

References

1. Hensel M, Holden DW (1996) Molecular genetic approaches for the study of virulence in both pathogenic bacteria and fungi. Microbiology 142:1049–1058
2. Reyrat JM, Pelicic V, Gicquel B, Rappuoli R (1998) Counterselectable markers: untapped tools for bacterial genetics and pathogenesis. Infect Immun 66:4011–4017
3. Stibitz S (1994) Use of conditionally counterselectable suicide vectors for allelic exchange. Methods Enzymol 235:458–465
4. Chakraborty T, Leimeister-Wachter M, Domann E, Hartl M, Goebel W, Nichterlein T, Notermans S (1992) Coordinate regulation of virulence genes in *Listeria monocytogenes* requires the product of the *prfA* gene. J Bacteriol 174:568–574
5. Wuenscher MD, Kohler S, Goebel W, Chakraborty T (1991) Gene disruption by plasmid integration in *Listeria monocytogenes*: insertional inactivation of the listeriolysin determinant *lisA*. Mol Gen Genet 228:177–182

6. Arnaud M, Chastanet A, Debarbouille M (2004) New vector for efficient allelic replacement in naturally nontransformable, low-GC-content, Gram-positive bacteria. Appl Environ Microbiol 70:6887–6891
7. Bae T, Schneewind O (2006) Allelic replacement in *Staphylococcus aureus* with inducible counter-selection. Plasmid 55:58–63
8. Monk IR, Gahan CG, Hill C (2008) Tools for functional postgenomic analysis of *Listeria monocytogenes*. Appl Environ Microbiol 74:3921–3934
9. Abdelhamed H, Lawrence ML, Karsi A (2015) A novel suicide plasmid for efficient gene mutation in *Listeria monocytogenes*. Plasmid 81:1–8
10. Monk IR, Shah IM, Xu M, Tan MW, Foster TJ (2012) Transforming the untransformable: application of direct transformation to manipulate genetically *Staphylococcus aureus* and *Staphylococcus epidermidis*. MBio 3(2): e00277. https://doi.org/10.1128/mBio.00277-11
11. Horton RM, Cai ZL, Ho SN, Pease LR (1990) Gene splicing by overlap extension: tailor-made genes using the polymerase chain reaction. Biotechniques 8:528–535
12. Gellin BG, Broome CV (1989) Listeriosis. JAMA 261:1313–1320
13. Sambrook J, Russell DW (1989) Molecular cloning: a laboratory manual. Cold Spring Harbor Laboratory, New York

Chapter 16

Whole-Genome Identification and Characterization of Bacterial Insertion Sequences Using Bioinformatic Tools

Kody A. Bassett, Melanie R. Mormile, and Ronald L. Frank

Abstract

Insertion sequences are small mobile regions of DNA (transposable elements) found primarily in prokaryotes. The identification of insertion sequences in bacteria is a growing field of study because of their applications in evolution, genetics, and medicine. One of the first steps in characterizing the insertion sequences found in an organism is to perform a genome-wide survey to identify all insertion sequences using in silico methods. This includes a thorough scan of the genome to locate all copies of different families of insertion sequences and the identification of the key characteristics of each element. The results provide an extensive catalog of the insertion sequences which can be used to further other analyses or manipulation of the genome.

Key words Extremophilic bacteria, Genome survey, Insertion sequences, Transposable elements

1 Introduction

Transposable elements, mobile DNA sequences, are the most abundant and diverse sequences found in nature [1]. Insertion sequences in bacteria belong to the class II transposable elements [2] and are complex in nature. In 1998 there were only about 500 unique insertion sequences identified and characterized [3]. As of 2015 the ISFinder database [4], the largest collection of insertion sequences, contained over 4600 different insertion sequences [5]. The number of unique sequences is continually growing as more and more species of bacteria are discovered and their genomes sequenced.

Identifying and classifying insertion sequences requires multiple steps. The first step, though tedious and time consuming, is important in ensuring that the entire genome is surveyed. The initial scan of the genome provides an estimate of the location for the open reading frame (ORF) for each element. From this starting point it is possible to identify the length of the insertion sequence, the terminal inverted repeats, the target site duplication, and the

Steven C. Ricke et al. (eds.), *Microbial Transposon Mutagenesis: Protocols and Applications*, Methods in Molecular Biology, vol. 2016, https://doi.org/10.1007/978-1-4939-9570-7_16,

actual coding region of the transposase and accessory proteins. The length is determined by using BLAST [6], Basic Local Alignment Search Tool, to align copies of the element. From the alignments it is possible identify the ends of the elements relative to a reference sequence. Terminal inverted repeats (IR) are characteristics of insertion sequences and serve as the location for binding of the transposase enzyme. IRs are identified by comparing the beginning bases of the element to the reverse complement of the last bases of the element. Also a characteristic of insertion sequence mobility is a target site duplication created by the staggered cuts made by the transposase enzyme at the point of insertion. These sequences are perfect direct repeats flanking both ends of the element. Finally, to identify the ORF(s) that encode the transposase enzyme various translation tools, such as EMBOSS Sixpack or Transeq [7] freely available on the online, can be employed.

In this chapter we outline the in silico procedure for surveying and identifying the insertion sequences found in a bacterial genome.

2 Materials

The methods outlined in this chapter are in silico and done on a computer using several open-source software programs available online.

2.1 Estimating Locations of Insertion Sequences in a Genome

1. ISFinder website.
2. Microsoft Excel or similar spreadsheet program.
3. National Center for Biotechnology Information (NCBI) genome database.

2.2 Element End Identification

1. Genome browser such as Argo Genome Browser at the Broad Institute.
2. NCBI BLAST.

2.3 Inverted Repeats Identification

1. Microsoft Word.
2. Sequence manipulation program for generating reverse complements such as "Reverse Complement" from bioinformatics.org.

2.4 Direct Repeats Identification

1. Genome browser such as Argo Genome Browser at the Broad Institute.

2.5 Open Reading Frame Identification

1. EMBOSS Transeq.
2. EMBOSS Sixpack.

3 Methods

3.1 Estimating Locations of Insertion Sequences in a Genome

To survey the genome for insertion sequences, a set of programs and procedures can be used simultaneously to predict the location of insertion sequences in the genome. These procedures are adapted from ISsaga (Insertion Sequence Semi-automatic Genome Annotation) program developed at the Laboratoire de Microbiologie et Genetique Moleculaire in Toulouse, France. We recommend, in addition to the steps outlined in Subheading 3.1, that researchers submit their bacterial genome to the ISsaga program [8].

3.1.1 BLAST of the Genome to the ISFinder Database

1. Using BLASTX on the ISFinder website, use the genome of interest as the query and leave the alignment view, word size, *E*-value, and gap open options as default. Save the results to be used later.
2. Perform the same BLASTX as above; however, in the alignment view options choose comma-separated values.
3. Copy and paste the comma-separated values into a blank text file and change the extension to .csv (e.g., BLASTX results.csv).
4. Open the csv file as a spreadsheet and sort the list by the column containing the left end of the query. Add an IS family column into the spreadsheet.
5. Using the BLASTX from **steps 1** and **2**, match the ORFS between the two BLAST output documents and add the appropriate IS family tag to each row in the spreadsheet.
6. The BLASTX contains many copies of each insertion sequence based on which organism they were found in. Use the approximated ORFs to group the same insertion sequences together (*see* **Note 1**).

3.1.2 Using Keywords to Identify Possible Insertion Sequences

1. Access the features table of the organism of interest, using a genome database.
2. Refer to the list of keywords in **Note 2**. This list was created using protein annotations associated with known insertion sequences.
3. Search the features table using the keywords and record the locus tag, ORF location, and keyword of any matches in a spreadsheet.
4. Using the approximated ORF compare the list created using the keywords to the BLAST in Subheading 3.1.1 to identify insertion sequences and assign locus tags to insertion sequences.

3.1.3 Using Insertion Sequences from Taxonomic Relatives to Identify Potential Insertion Sequences

1. Identify the taxonomic ranks for the organism of study, *see* **Note 4**.
2. Starting with genus, do a Boolean search in the GenBank database using the rank and one of the IS families in the provided list (*see* **Note 3** for list).
3. If there are matches, search the features table of each species for all instances of that insertion sequence, and do a TBLASTN of those insertion sequences to the species of interest. Skip **step 4**.
4. If there are no matches at the genus rank, work up the taxonomic ranks until there are matches, *see* **Note 5**. Then perform **step 3** for as many of the species as possible and deemed necessary to produce a more exhaustive and complete list.
5. Record the BLAST results in a table including the IS family that was used and the locations in the genome of the subject.
6. Repeat these steps for every insertion sequence family in the provided list of IS.

3.2 Element End Identification

The ends of the insertion sequences are unknown and must be determined. Element ends are typically within a few hundred nucleotides of the start and stop of any protein coding sequences contained within the element. An extended sequence that includes 500 nucleotides upstream and downstream of the open reading frame(s) should contain the actual ends of the element. This sequence will serve as a query to BLAST (Basic Local Alignment Search Tool) the entire genome of the organism. Most alignments will be other elements in the same insertion sequence family and will be identified as isoforms or partials in subsequent steps. The alignments can be used to determine the actual ends of the element.

1. A reference element must be chosen that is representative of the elements in that family. Refer to **Notes 6–10** for choosing a reference element.
2. Subtract 500 bases from the start of the approximated ORF (e.g., if the ORF started at base 114,190 of the genome, the extended left end of the element is 113,690) (Fig. 1).

Fig. 1 The approximated ORF, acquired from Subheading 3.1, for the element is in black. The red arrows show the element minus 500 bases on the left and plus 500 bases on the right

3. Add 500 bases to the stop of the approximated ORF (e.g., if the ORF stopped at base 115,371 of the genome, the extended right end of the element is 115,871).
4. Use a genome browser to export the extended element as a text file.
5. Use the BLAST tool to perform a blastn, using the default parameters, to compare the extended element to the entire genome of the organism being annotated.
6. Label the first alignment as 1, the second 2, etc. until all alignments are labeled. (Note: alignment 1 is an alignment to the query itself).
7. Align the first 15 bases of the full size hits, excluding hit 1, to see similarities between them. Choose the closest base to the start that all the hits had in common as the start of the element. Repeat this with the end of the element, but use the last ten bases.
8. Use hit 1 to determine which base the element starts and stops at in the genome (Figs. 2 and 3).

3.3 Inverted Repeats Identification

Inverted repeats are generally located within 50 nucleotides of each end of the element (Fig. 4). Inverted repeats play a significant role in the recognition of the element by the transposase and are required for transposition activity. The inverted repeats are irregularly imperfect, so identification is not trivial.

1. Extract the first 50 bases from the start of the element and paste into a word document using a nonproportional font such as Courier New for proper alignment.
2. Extract the last 50 bases of the element and convert it to the reverse complement.

```
Query  601     AAAAATTATATTTGAGATTATTATTGCTGATGGTAAATAAGTAAATAAATTCATAAAAAC  660
               ||||||||||||||||||||||||||||||||||||||||||||||||||||||||||||
Sbjct  113990  AAAAATTATATTTGAGATTATTATTGCTGATGGTAAATAAGTAAATAAATTCATAAAAAC  114049

Query  661     CTCCATAATAAATAACGGTGATTTAGTAGTGTACAATAAGTTGTGTAAATAGAtttttt  720
               ||||||||||||||||||||||||||||||||||||||||||||||||||||||||||||
Sbjct  114050  CTCCATAATAAATAACGGTGATTTAGTAGTGTACAATAAGTTGTGTAAATAGATTTTTT  114109

Query  721     CAATaaaaaaaGAGGAATCCTTCTTTGAATGGTTGAAATATTTATTAGCGAAATAAACTC  780
               ||||||||||||||||||||||||||||||||||||||||||||||||||||||||||||
Sbjct  114110  CAATAAAAAAAGAGGAATCCTTCTTTGAATGGTTGAAATATTTATTAGCGAAATAAACTC  114169
```

Fig. 2 Highlighted is the common base start in the query (reference element) that corresponded to base 114,074 in the subject (the genome of the organism of interest) in hit 1

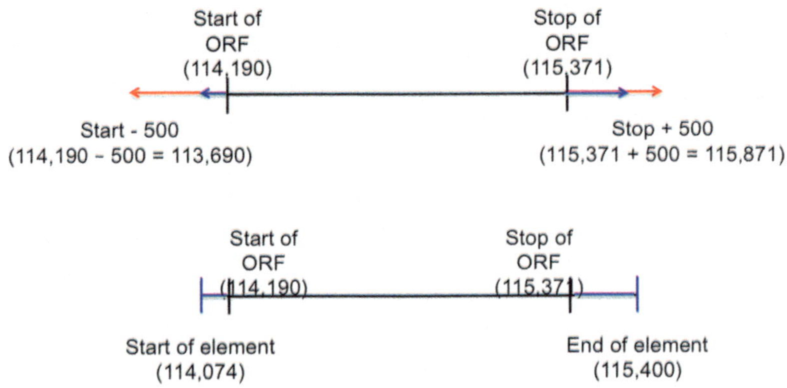

Fig. 3 The top element shows the actual ends of the element, determined by the BLAST, in blue, relative to the extended element shown in red. In the bottom sequence the red arrows have been removed and the blue lines are showing the actual ends of the element compared to the ORF in black

Fig. 4 The red arrows represent the inverted repeats. The insertion sequence is in black, and the lines represent the ends of the element. The arrows are facing each other because they are reverse complements of each other

19/28
AGTAGTGTACAATAAGTTGTGTAAATAGATTTTTTTCAATAAAAAAAGAG
GATAACGTCAAGAATCTTGTGTAAATAAATCAGTATCGTTCTTTAAATAG

Fig. 5 The top sequence is the first 50 bases of the element and the bottom sequence is the reverse complement of the last 50 bases of the element. The highlighted bases are identical between the two strands. A score is assigned by taking the number of base matches over the number of bases

3. Arrange the reverse complement of the last 50 bases under the first 50 bases. Highlight any matches between the two sequences (Fig. 5).
4. Choose the best score of matches to be the inverted repeats for the element; for example, in Fig. 5 the best score for this element is 19/28, even though there is a total of 28/50 matches, or a score of 0.56. This score is smaller than 19/28, which is a score of 0.68 (Fig. 4). Refer to **Note 11**.

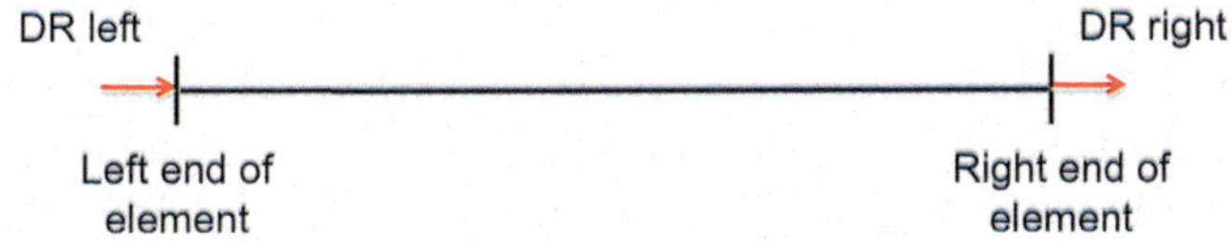

Fig. 6 The red arrows represent the direct repeats. The insertion sequence is in black, and the lines represent the ends of the element. The arrows are facing the same direction because they are exact copies of each other flanking the ends of the element

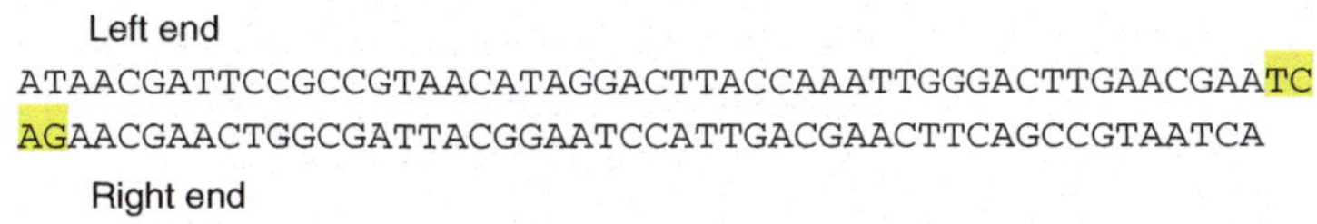

Fig. 7 Highlighted on top are the first two bases of the insertion sequence and highlighted on the bottom are the last two bases of the sequence

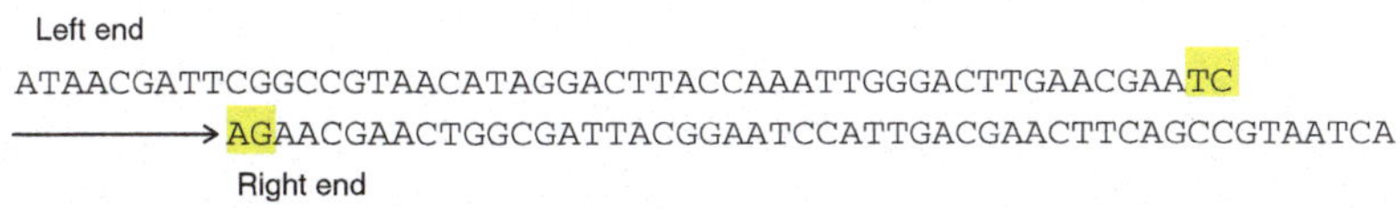

Fig. 8 The arrow shows the sliding of the right end of the element closer to the left end to identify the direct repeat

3.4 Direct Repeats Identification

Direct repeats are perfect matches located immediately adjacent to both ends of the element (Fig. 6). The direct repeats are a result of a duplication of the insertion site during the process of transposition and are also known as target site duplications (TSD). The direct repeats vary in size based on the insertion sequence family. Together, the inverted repeats and direct repeats are used to verify that the correct ends were chosen in Subheading 3.2.

1. Use a genome browser to export a sequence extended by 50 bases in both directions from the ends of the element.
2. Delete the element, leaving the extended 50 bases and 2 bases from the ends of the element.
3. Highlight the two bases from the ends of the element and align them (Fig. 7).
4. Slide the right end, one base at a time to the right to bring the highlighted ends of the element closer. Check for a match after each movement (Fig. 8).
5. A perfect match is the direct repeat (Fig. 9). If a perfect match is not identified the ends may not be correct, refer to **Notes 12–14** if this occurs.

```
                                                 DR
Left end
ATAACGATTCCGCCGTAACATAGGACTTACCAAATTGGGACTTGAACGAATC
                                          AGAACGAACTGGCGATTACGGAATCCAT
                                           Right end
```

Fig. 9 Highlighted in green is the perfectly matched direct repeat flanking both ends of the insertion sequence

3.5 Open Reading Frame Identification

Insertion sequences harbor genes that encode the transposase proteins that copy or move the element. Some families of elements carry additional genes as well. After the element ends have been determined and verified by inverted and direct repeat identification, a search for open reading frames (ORFs) is necessary. A large number of ORF identification programs exist with varying degrees of sophistication. We will describe a method that employs simple translation.

1. Use EMBOSS Transeq to translate the element in all three forward frames. If the insertion sequence is on the minus strand refer to **Note 15**.
2. Identify a long uninterrupted sequence of amino acids; an approximate size should be acquired in Subheading 3.1.
3. Typically the first methionine of the ORF is assigned as the start for the encoded protein. The stop at the end of the long sequence of amino acids is the end of the coding region. Refer to **Note 16** for special cases.
4. Align the translated ORF to the DNA sequence of the element using EMBOSS Sixpack to determine the locations of the start and stop of the ORF within the element.

4 Notes

1. The results from this BLAST will be very large and will contain a high amount of variation. It is recommended that BLASTX results be compared with the results from Subheadings 3.1.2 and 3.1.3 to identify insertion sequences.
2. List of keywords: AAA family ATPase, ATP—binding protein, ATPase AAA, chromosomal replication initiator protein DnaA, DDE domain—containing protein, DDE transposase, DNA—binding response regulator, endonuclease family protein, HD domain—containing protein, hypothetical protein, integrase, any IS family transposase. MBL fold metallo—hydrolase, resolvase, and transposase. The compiled list was created using a sample case. Therefore, the list is not all-inclusive, modify and add your own keywords to the list if you feel they could be associated with an insertion sequence.

3. List of discovered insertion sequence families: IS1, IS3, IS4, IS5, IS6, IS21, IS30, IS66, IS91, IS110, IS200/IS605, IS256, IS481, IS607, IS630, IS701, IS982, IS1182, IS1380, IS1595, IS1634, ISAs1, ISAzo13, ISH3, ISL3, Tn3, ISH6, ISNCY, ISLre2.
4. This step is the most time-consuming of the steps and is used as a last resort to find insertion sequences that may have been missed in previous steps. It may be useful to use this step for insertion sequence families that were not identified in previous steps.
5. Not every insertion sequence family will be represented in every organism. If you get to the higher taxonomic ranks (e.g., phylum or class) before identifying a match, there is a good chance that the insertion sequence does not exist in the organism of study.
6. For convenience and if possible, the reference element should be on the positive strand. Though some families may only have elements on the minus strand.
7. There is a chance that the reference element chosen is not representative of the other elements in the family (has a frame-shift causing an early or later stop, the element was interrupted by another element, etc.). If this happens, go back and choose a different reference.
8. There can be multiple reference elements in a family. If the BLAST of two amino acid sequences is less than 98% positives, they are considered different elements.
9. If the reference element is on the minus strand, the ORF starts on the right and stops on the left. Therefore, when extending the ends of the element, 500 bases need to be subtracted from the stop and 500 bases added to the start (Fig. 10).
10. When exporting an element that is on the minus strand, the genome browser will give the sequence for the positive strand. The reverse complement is taken to convert the sequence into the minus strand.

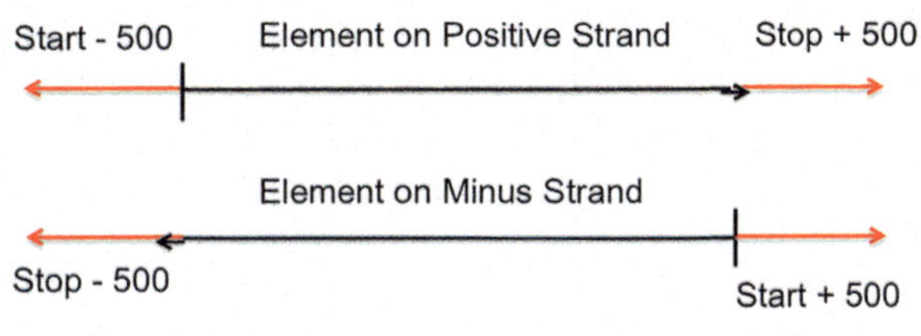

Fig. 10 The top element is on the positive strand and starts on the left and ends on the right. The bottom element is on the minus strand and runs in the opposite direction. Therefore, 500 bases are subtracted from the stop and 500 bases are added to the start

11. If repositioning either the top or bottom strand one or two bases results in a significantly higher score, it may be necessary to revisit Subheading 3.2 to revise the element ends before continuing.
12. If a perfect match is not found in between the highlighted region, the ends may not be correct. To check, unhighlight one base from either end and redo the sliding to check for a perfect match or revisit Subheading 3.2.
13. Some individual elements within a family may not have a DR. This could be an indicator of recombination between elements.
14. Direct repeat size can vary greatly between elements, ranging from 2 to 14.
15. If identifying the ORF of an element on the minus strand you should be looking at the reverse three frames or frames 4–6. If you first convert the element to its reverse complement, then you will be looking at the forward three frames as described in the methods.
16. Some families may have two or more ORFs and can be found in the literature along with what type of ORFs (e.g., overlapping ORFs requiring a frameshift or auxiliary or helper proteins).

References

1. Aziz R, Breitbart M, Edwards R (2010) Transposases are the most abundant, most ubiquitous genes in nature. Nucleic Acids Res 38:4207–4217
2. Han Y, Qin S, Wessler SR (2013) Comparison of class 2 transposable elements at superfamily resolution reveals conserved and distinct features in cereal grass genomes. BMC Genomics 14:71. https://doi.org/10.1186/1471-2164-14-71
3. Chandler JM a M (1998) Insertion sequences. Microbiol Mol Biol Rev 62(3):725–774
4. Siguier P, Perochon J, Lestrade L, Mahillon J, Chandler M (2006) ISFinder: the reference centre for bacterial insertion sequences. Nucleic Acids Res 34:D32–D36
5. Siguier P, Gourbeyre E, Varani A, Ton-Hoang B, Chandler M (2015) Everyman's guide to bacterial insertion sequences. Microbiol Spectr 3(2): MDNA3-0030-2014
6. Altschul SF, Madden TL, Schäffer AA, Zhang J, Zhang Z, Miller W, Lipman DJ (1997) Gapped BLAST and PSI-BLAST: A new generation of protein database search programs. Nucleic Acids Res 25:3389–3402
7. Rice P, Longden I, Bleasby A (2000) EMBOSS: the European molecular biology open software suite. Trends Genet 16(6):276–277
8. ISsaga: A platform for identification and semi-automatic annotation of prokaryotic insertion sequences. http://issaga.biotoul.fr

Index

Steven C. Ricke et al. (eds.), *Microbial Transposon Mutagenesis: Protocols and Applications*, Methods in Molecular Biology, vol. 2016, https://doi.org/10.1007/978-1-4939-9570-7, © Springer Science+Business Media, LLC, part of Springer Nature 2019

S

T

V

W

Printed by Printforce, the Netherlands